2

D0586084

AN ILLUSTRATED HISTORY OF
FORK-LIFT TRUCKS

AN ILLUSTRATED HISTORY OF
FORK-LIFT TRUCKS

HINTON J. SHERYN

Ian Allan
PUBLISHING

First published 2000

ISBN 0 7110 2736 6

© Hinton J. Sheryn 2000

Published by Ian Allan Publishing

an imprint of Ian Allan Publishing Ltd, Terminal House,
Shepperton, Surrey TW17 8AS.
Printed by Ian Allan Printing Ltd, Riverdene Business
Park, Hersham, Surrey KT12 4RG.

Code: 0011/B

Title page:
**A SkyTrak 6036 draws on its considerable power
and stability to negotiate thick mud. It will use its
telescopic capabilities to place the load of
shuttering boards close to where they are
required.** *SkyTrak International*

Half-title page:
**One of the range of rough-terrain fork-lifts
produced during the 1950s by Matbro, seen here
lifting a reel of Kraft paper on a dump near
London, working for Bowater Fibre-board
Containers Ltd, c1957.** *Ian Allan Library*

ACKNOWLEDGEMENTS
Jonathan Field (Adfield-Harvey); Joan Cole (Barlow
Handling); Louise England (Boss UK Holding Co Ltd); Jim
Brindley (BT Rolatruc); Kevin Hudson (BT-Rolatruc /
Clark UK); Tom Nankervis (Claas); Giuseppe Lilla (Cesab
UK Ltd); Christeen Jarrett (Cesab UK Ltd); Barry Lea
(Fork-Lift Truck Association); Jozette MacLean (4 Point Lift
Systems); Georgina Lankester (GL PR Services); Anthony
Davy (Hyster Europe); John Palmer (JCB); Jutta Carlyle
(Jungheinrich (GB) Ltd); Lotta Sundstrom (Kalmar
Industries); Emma Clitheroe (Lansing-Linde); Bart Meertens
(Lisman Vorkheftrucks BV); Seamus Merchan (Moffett
Engineering Ltd); Helen Bishop (Narrow Aisle Ltd);
P. Wooldridge (Narrow Aisle Ltd); Cynthia L. Kruell
(Omniquip Lull International Inc); Bob Pooler (Pooler-LMT
Ltd); Steve Harris (S. Harris Materials Handling); Dick
Shelley (Sales & Busines Consultants); Harry Hinder
(formerly of Stacatruc / Clark); Tony Fretwell (Terrain
Forklifts); Miles Griffin (Toyota Industrial Equipment Ltd);
F. L. Brown (Translift); Peter Grant (US Eurolink Associates);
Grahame Miller (VME); Les West (West Lancs Forklift);
Andy Woolfenden (for the use of his photograph collection);
Peter Waller (Ian Allan Publishing); Mavis Pearson (for her
help, throughout).

CONTENTS

INTRODUCTION

From the beginning of the 1930s, pallet-handling trucks have handled vast quantities of the world's materials. Almost every commodity known to man has been moved at some stage by these types of mechanical handler. The list is endless, but includes timber, iron and steel, chemicals, oil, raw and processed foods (both for humans and for animals), factory products from glass to the automotive industry (including parts), plastics, stationery and paper products, precious and semi-precious metals, consumer products and electrical goods of every kind, and materials for the building and construction industry. Only the crane can rival the fork-lift truck's claim to be the prime mover of the world's products. In recent years, however, the fork-lift truck has taken over many of the tasks once performed by cranes on construction sites, in factories, warehouses and timber yards, and at docksides and airfields.

It was the railways and their increased use for carrying goods, mail and parcels that brought about the use of small, specially-built wheel tractors to pull one, two or more trailers from the warehouses out on to the platforms, where they were off-loaded onto the trains. Throughout this book I hope to trace the development of these tractor units and fork-lift trucks of a wide variety of types, along with that of their manufacturers.

Aside from powered lift trucks, one development which is sold by the thousand to factories, large stores, retail outlets and warehouses is the manually-operated truck mounted on steel roller wheels which, while working on a flat, hard concrete floor, can easily be manoeuvred by hand. The handles are used for both steering and for pumping up the small hydraulic cylinder, which raises or lowers the loaded pallet-forks.

Below:
Liverpool Street station in London in 1943 — typical of all the main-line platforms in the days when goods as well as passengers were moved. Note the numbers of trolleys and at least one little powered platform-tractor. *Ian Allan Library*

Finchley Road in north London is the location of this goods yard, where hand-operated pallet trucks are essential in the movement of large quantities of goods.
Ian Allan Library

Many manufacturers began producing these in the early 1940s, as the standardisation of pallet sizes and uses became a reality. Pallets are commonly constructed from wood, steel or durable plastic. Loads of up to 3,000kg (6,700lb) are regularly moved using the manual-type lifter. More and more options for the wheels on these types of machine are becoming available, such as nylon, polyurethane, vulkollan, steel or rubber, dependent on the specifications required by the purchasers.

Besides those normally working at just above ground level, some manufacturers now offer a high-lift version of manually-operated trucks enabling the operator to load low-level shelving without having to lift heavy loads from floor level. This should help eliminate the all-too-common backstrain experienced in industry while personnel are daily asked to perform lifting tasks which can have a harmful effect on the person, his employer and the medical system. Modern manual trucks come with a choice of manual or electric lift capabilities.

Below:
Many wheel-loaders, including those made by Muir-Hill, Allis-Chalmers, JCB and Terex have been used at airports to handle supplies of humanitarian aid to wherever it is needed in the world. This machine is loading supplies for victims of the Bosnian conflict. *Author*

Larger, powered fork-lift trucks which are operated by a pedestrian come in a variety of configurations. With capacities of between around 800kg (1,800lb) and 2,000kg (4,500lb) they are available for shelf-stacking and moving loads around factory or warehouse floors. A system utilising a flip-down driver platform enables the operator to ride on the machine in a standing position, while other types have an enclosed-operator standing position. Further up the scale the operator has a seat provided for both comfort and speed of operation. Certain models of such machines have extra-long forks, enabling them to move two pallets horizontally at a time.

Why, you might ask, have I included wheel-loaders into this book? Although these are more commonly known to use a wide range of buckets to handle loose material — earth, minerals or rock — they do use forks to add to their versatility, enabling them to move timber, steel pipes, electric cables or a whole host of building materials.

Although I have made every effort to identify all the machines in all of the photographs as accurately as I can, I cannot guarantee to be correct in my analysis on every occasion. Indeed, fork-lift trucks are no different from excavators, earthmovers or cranes, in that little effort has been made to remember their origins. One umbrella organisation destroyed hundreds of early photographs just weeks before I embarked on this book. They featured most types and sizes of fork-lift truck made (certainly in the UK) since the 1930s. On hearing that news I wondered if I had made a fatal error of judgement in going ahead with this title. However, it is the very lack of documentation of facts and photographs from this multi-billion-dollar industry that made me want to continue.

I hope you as the reader will be somewhat enlightened and enjoy this book.

Hinton J. Sheryn
June 2000

MANUFACTURERS

Allis-Chalmers

Allis-Chalmers of Matteson, Illinois, USA, was for many years one of the world's best-known manufacturers of construction and agricultural machinery, with wheeled and crawler tractors, tractor-shovels, bulldozers, scrapers, combine-harvesters etc; it also produced a wide range of fork-lift trucks, which included electric-rider lift trucks (models ACE 20, 25, 30, through to the 120), electric stand-up-end control trucks (models SC20, SC25, 30 and SC40), narrow-aisle trucks (models S520, 30, 40, SR20, 30 and SR40), electric order-selecting trucks (models ACOP15S, ACOP 25SL, ACOP 30C), electric walkie-trucks (low-lift models LWF45, LWF60, LWP45, LWP60, LWA60, LWB45, LWB60 and ET700, and high-lift models HWC20, HWC30, HWS20, HWS30, HWS40, HWR20, HWR30 and HWR40), cushion-tyre lift trucks from 2,000lb to 12,000lb capacity (models ACC20 to ACC100C — 16 models in all), pneumatic-tyre lift trucks (from the ACP20 at 2,000lb to the ACP80 at 8,000lb — eight models in all), large-capacity pneumatic-tyre lift trucks from 10,000lb to 82,000lb capacities (ACP110 to FH820 — 19 models) and sideloaders (from the 3,500lb-capacity model S35 to the 81,000lb S810 — 13 models in all). In addition there were six models of the Weight Mate compact lift truck (from the 1,000lb-capacity ST2024 to the 3,000lb PT2524) running on either solid or pneumatic tyres.

The majority of the Allis-Chalmers line of earth-moving equipment was salvaged from a poor marketing base when in 1974 it was decided to enter into negotiations with Fiat of Italy. Thus was formed Fiat-Allis, a joint venture, with Fiat holding a 65% share and Allis-Chalmers the remaining 35%.

Above: **The Allis-Chalmers ACOP 30C, used as an order selector, transporter and right-angle stacker.** *Allis-Chalmers*

Overleaf: **An early Allis-Chalmers catalogue illustrating a selection of the company's fork-lift trucks, including pedestrian-controlled trucks and sideloaders, diesel- and electric-powered for warehouse and yard work.** *Allis-Chalmers*

ALLIS-CHALMERS

Bonser

Below:
A couple of Bonser heavy-duty lift trucks handling timber at Liverpool docks.
Ian Allan Library

Boss

The name 'Boss' has been synonymous with the development of both the British and the international lift-truck industry during the second half of the 20th century, the company having grown over the last 40 years to become one of the world's leading manufacturers. Its successful track record over the years has derived from a unique amalgam of British creativity, innovatory designs, advanced engineering skills, quality products, marketing flair and a customer-support reputation which is second to none.

In 1957 (four years after Dr Ing Friedrich Jungheinrich laid the foundations of the present parent group in Hamburg), Neville and Trevor Bowman-Shaw spotted a need for a side-loading truck capable of dealing with long loads in ports, terminals, steel stockholding centres, timber yards and similar operations. Under the name of 'Boss' (a contraction of 'Bowman-Shaws'), they therefore combined with Lancers Machinery Ltd to manufacture and market specialist materials-handling equipment. Designed by

John Kinloss, the first all-British sidelift truck was built in shared works in Slough in 1958. Additional sidelift models joined the range in 1959, when the company relocated its manufacturing, spares and service operations to a new, purpose-built factory in Leighton Buzzard, Bedfordshire. This marked the beginning of the Boss organisation, which was to expand rapidly over the next four decades to become a leading player on the international handling stage.

By 1965, frontlift machines were able to cover capacities of 2 to 25 tons, while the sidelift trucks were available from 1.5 to 12 tons. In 1966 two series of large sidelifts, of 25- and 37-ton capacity, entered the range. Sales and service subsidiaries were established in Ireland and Austria in 1967. With the growth in container traffic, 25-ton and 35-ton-capacity sidelift container-handling versions, able to handle 20, 30 and 40ft-long freight containers, became available in 1968. The first internal-combustion-engined frontlift trucks — the original B series — were launched in 1969 in capacities of 10-12 tons. In 1981 the G series of frontlift trucks was launched, which eventually covered 3.5- to 50-ton capacities, including container handlers and Ro-Ro (roll on, roll off) models.

In 1983 Lancer Boss acquired Steinboch GmbH of Moosburg, which at that time ranked as Germany's third-largest truck manufacturer, and which was known thenceforth as Steinboch Boss.

On 14 April 1994 Steinboch Boss was acquired by Jungheinrich AG, which on 5 May of the same year concluded negotiations to purchase the UK operation of Lancer Boss, the latter trading from then on as the Boss Group Ltd. Thus Jungheinrich AG now includes Boss (UK), Steinboch (Germany) and MIC (France).

The group's products have been manufactured under licence in India and in Yugoslavia, with agreements being signed with Komatsu and Nissan for the production of certain models here in the United Kingdom of these Japanese trucks. New models were continually added to the range throughout the 1980s and 1990s covering every type of truck.

Manufacturing facilities extend to Norderstadt on the outskirts of Hamburg, which concentrates its efforts on reach trucks; Moosburg in Bavaria, specialising in three- and four-wheeled electric counterbalanced trucks, and very-narrow-aisle (VNA) warehouse trucks; Argentan in Normandy, which produces pallet trucks (at a rate of 160,000 per year) and powered pedestrian trucks; and Leighton Buzzard in Bedfordshire, where the products are internal-combustion-engined fork-lifts, heavy-duty frontlifts, sideloaders and container-handling trucks, with capacities ranging from 1½ to 50 tons. Another important role for Boss at Leighton Buzzard is the production of the volume range of internal-combustion-engined trucks previously supplied to Jungheinrich AG during a 14-year marketing agreement with NACCO/Yale.

Far left:
This Lancer 2500-series Sideloader is the type of machine normally seen operating in timber yards, due to its ability to reach out and stack lengths of timber. Here, however, it is able to demonstrate its capabilities as a container stacker, going all-out for three-high with ease. *Ian Allan Library*

Above:
A Lancer 2500-series Sideloader lifts a double-decker bus — its lifting capacity was 55,000lb. Lancer's Machinery Ltd of Leighton Buzzard, Bedfordshire, went on to become part of the Boss Group. *Ian Allan Library*

Above:
The mighty Boss D-series is able to unload or load an articulated trailer in one go.
Ian Allan Library

Below:
Yet another Boss D-series, stacking Lancashire flats at the Ellerman Wilson Line terminal in Hull.
Ian Allan Library

Above:
'Cold without a cab, mate, innit?' It is little wonder that many of the old dockside cranes find themselves redundant, with big fellas like this Boss about.
Ian Allan Library

Right:
A novel idea this: a Boss A25/48 recovery vehicle (fork-lift truck to you and me) with a lifting capacity of 25,000lb at 48in load centre, operating in the new Tyne Tunnel in December 1967.
Ian Allan Library

Below:
A 23½-ton-capacity Lancer Boss fork-lift, powered by a Perkins V8 engine. Built c1975, it was owned by Greater Manchester Waste Ltd and is seen here working at the Holiday Moss landfill site.
Andy Woolfenden

Bottom:
Pictured at an ex-Army sales company's yard, this United Nations Lancer Boss fork-lift is capable of lifting containers three-high, using a spreader bar which lifts the container from the top.
Andy Woolfenden

Caterpillar

Bottom:
A Caterpillar V550 25-ton-capacity truck, powered by a Caterpillar V8 engine, loading containers for Greater Manchester Waste Ltd at Salford treatment plant. This machine is kept very busy, loading road vehicles and railway trains. *Andy Woolfenden*

Below:
A Caterpillar AH52 24½-ton-capacity chain-driven fork-lift truck loading waste containers on to dump-trucks which are purpose-built to carry these large objects. The containers have been transported to the site by rail, and will now be taken to the tipping face at Wimpey Waste's Parbold Hill landfill site at Appley Bridge, near Wigan. This AH52 was one of three working at the site, all of which had already completed 10 years' work when this photograph was taken in 1991. *Andy Woolfenden*

Above:
Pictured in 1997 at Holiday Moss landfill site, this Caterpillar V620 is one of two operated by UK Waste Management Ltd. It can lift 28 tons on its 8ft forks, and is fitted with a weighing system, designed to weigh each waste container when lifted. *Andy Woolfenden*

Below:
The twin of the machine shown above, this Caterpillar V620 is working at Roxby landfill site at Scunthorpe. It is kept busy unloading/loading railway trains laden with waste containers, and is fitted with closed-circuit television cameras front and rear to assist the operator. It weighs around 42 tons. *Andy Woolfenden*

Cesab

Clark

Clark is one of the many manufacturers of fork-lift trucks which can lay claim to being one of the true pioneers in the industry. Eugene B. Clark and his business partner founded the George B. Rich Manufacturing Co in Chicago in 1910. They initially produced tools and moulds, later supplying steel wheels and axles to the growing automotive industry.

In 1916 the company began trading under the Clark Equipment Co banner, and in 1917 developed the Tructractor, a three-wheeled, gas-driven vehicle for transportation within the company. This was the true forerunner of the industrial truck, but only one was initially produced. Visitors to the factory, however, showed considerable interest in the Tructractor, such that Clark produced further models, finally opening a separate factory for vehicle production.

In 1921 the production range of the new business was expanded considerably as Clark presented the Truclift — the first hydraulic platform truck — to the world. Between 1924 and 1927 other new products were unveiled, including the Duat, the first gas-driven tractor with a mechanical lifting mechanism; in 1928

Above:
The fork-lift truck industry is turning over billions of pounds annually for the manufacturers, their dealers and distributors, and not least for the used fork truck business. Some companies are happy to trade with a stock of just a few machines, while others are capable of holding hundreds or indeed thousands of trucks. One company, Lisman Vorkheftrucks BV of the Netherlands, holds over 1000 trucks of all makes and sizes; via integrated computer links with dealers and clients all over Europe, it is able to locate the make, model, year and power unit of the customer's choice. The company supplied this Clark C500-Y-100 built in 1982.
Lisman Vorkheftrucks BV

Below:
The very large transfer warehouse at Paddock Wood was keeping at least four fork-lift trucks busy in December 1974. The trucks appear to have been built by Clark, and employed a variety of power units — electric, LPG and diesel.
Ian Allan Library

Clark developed the Tructier, with hydraulic lifting device and front-wheel-drive — basically the forerunner of later fork-lift trucks.

At the outbreak of World War 2 around 90% of all tractive machines and lift trucks used for materials transport in the USA were Clark machines. In 1942 the company developed its first electrically-driven industrial trucks, and a year later production reached a pinnacle with 23,000 lift trucks.

In 1945, Clark lift trucks began arriving on British shores and in Europe generally. A repair facility was provided for Clark lift trucks by Hugo Stinnes in Mulheim in the Ruhr in 1947, where in 1952 a new manufacturing plant was opened to produce Clark trucks under licence. This plant was expanded in 1976/7 at a cost of around $50 million, with capacity for about 12,000 lift trucks per year.

A range of 22 M-series models was being produced by 1985 when Clark was acclaimed as manufacturer of the Lift Truck of the Year. In 1992, Clark sold its material-handling company to Terex Corporation, which manufactures everything from giant dump trucks, loaders, scrapers, excavators, cranes, crushing and screening plants to a whole host of associated products. Terex is growing rapidly by acquisition, and is believed to have its eyes on one or two major manufacturers in an effort to expand still further.

Currently the BT organisation markets Clark lift trucks in the UK, along with its own BT Rotatruc range of machines.

Conveyancer

Conveyancer lift trucks were produced by Rubery Owen of Warrington for many years, and the range included the following:

RE2-24 and RE3-24	battery-electric reach trucks with capacities of 2,000lb at 24in (RE2-24) or 3,000lb at 24in (RE3-24)
E2-20, E2-24 and E22-24	battery-electric three-wheeled designs with capacities ranging from 2,000lb at 20in to 2,240lb at 24in
E3-24	battery-electric four-wheeled design
E4-34	battery-electric four-wheeled design
FE4-20 and FE45-20	flameproof electric four-wheeled design, suitable for operations where gases and vapours exist
E5-25	battery-electric
E6-24 and E7-24	battery-electric
D4-24/G4-24	diesel/petrol series III four-wheeled design
G4-24	low-pressure gas (LPG) four-wheeled design
TC4-24, TC5-24	diesel/petrol series Vb four-wheeled design
FTC5-24	flameproof diesel series V four-wheeled design
TC7-24	diesel/petrol series Vb four-wheeled design
TC6-24 and TC7-24	yard model with double-drive wheels

Right:
As already stated, fork-lifts can handle almost anything. Here a Conveyancer is moving wire (most likely copper wire, as used in the manufacture of electric motors). Seen on 24 July 1948.
Ian Allan Library

Right:
A pallet laden with cast-iron pipes ready for delivery with help from a Conveyancer fork truck in 1948. *Ian Allan Library*

Below:
A Commer articulated lorry is seen entering the hold of the *Empire Gaelic*, bound for Belfast, in Preston in June 1955. *Ian Allan Library*

Also available was the Overlander, Conveyancer's rough-terrain vertical-mast fork-lift truck.

Other machines were built under licence, or following the acquisition of subsidiary companies. These included the following:

Conveyancer-Raymond E4-RT electric reach fork truck;
Conveyancer-Raymond E4R-SW electric reach fork truck;
Conveyancer-Raymond WL 45F powered pallet truck;
Conveyancer-Scott fixed-platform truck (10cwt and 1-ton capacity);
Conveyancer-Scott platform tractor (2-ton and 3-ton capacity);
Conveyancer-Scott Rider industrial tractor (2-ton capacity);
Conveyencer-Scott low-load elevating-platform truck and low-load fixed-platform truck (1-ton and 2-ton capacity);
Conveyancer-Scott Electric Horse (high-pick-up model of 3-ton capacity);
EMI Robotug tractor;
Shorland TC10-24, TC21-24 heavy-duty front-loader fork trucks;
Shorland Series 20 and 21 Straddle Carriers — able to carry heavy loads including steel pipes, timber, steel;
Hopper loading units — fixed elevators able to hoist bins or trolleys containing chemicals, paints, dough or other products;
Lifting platforms.

Above:
On arrival in Belfast, the lorry's cargo is unloaded by a Conveyancer fork truck. *Ian Allan Library*

Right:
Electro-Hydraulics Ltd was the manufacturer of Conveyancer fork-lift trucks at the time this three-wheel version was produced. *Ian Allan Library*

Above:
This little Conveyancer electric fork truck was powered by Kathenode batteries, and was fitted with a special 6ft stacking-mast to handle the crates of milk at an Oxford dairy, as this 1950 view shows. *Ian Allan Library*

Left:
Anyone for tea? No shortage of tea chests in this warehouse, where a little Conveyancer has the responsibility of stacking and loading them for distribution. *Ian Allan Library*

Right:
Conveyancer was one of the early pioneers of the lift truck industry, certainly in the UK. The company was owned by Rubery-Owen in the 1960s and had a good share of the counter-balance truck market, 1- to 2-ton capacity and also had a licence agreement with Raymond to build the Raymond reach truck. Conveyancer was later bought by Coventry Climax; in management terms, it probably suffered from Lansing Bagnall's failure to buy Coventry Climax (due to intervention by the Monopolies Commission). Nevertheless, there are many Conveyancers still at work, often after 20 or more years of service. This little machine is handling what appear to be jute sacks in 1951.
Ian Allan Library

Right:
The Conveyancer seen here is loading cartons of powder milk on to a railway wagon, where the pallets are being stacked by a manually-operated Lansing-Bagnall lift truck, c1957. *Ian Allan Library*

Above:
Two Conveyancer trucks are seen unloading pallets from a British Railways articulated goods lorry prior to loading them on to railway trucks. The year is 1957.
Ian Allan Library

Left:
Conveyancer also entered the rough-terrain fork-lift truck market, as shown here in May 1963. These trucks were built at Warrington.
Ian Allan Library

Left:
Some of the most unusual machines developed for the materials-handling industry were the carriers, built by such companies as Hyster and Shorland, which were able to straddle, say, a stack of timber or iron pipes, lift them and transport them to wherever they were needed. These straddle-truck carriers were first introduced in the 1930s, and appear to have been the inspiration for the somewhat larger container handlers now familiar at the world's biggest container terminals.

The carrier seen here was built by Shorland, a company within the Conveyancer group. *Ian Allan Library*

Right:
The Conveyancer P50 LPG-powered four-wheel lift truck is a later model than those featured in the Conveyancer pocket range guide. Quite a sturdy little lift truck, many hundreds were still working at the end of the 1990s, long after they were built. *Author*

Left:
A close-up of the Conveyancer P50, with hood raised showing its engine, an LPG-powered unit. *Author*

Coventry Climax

A soldier of the 3rd Troop 37/38th Royal Buckinghamshire Hussars had returned from service in the Boer War. He was H. Pelham ·Lee, the son of a London architect, born in Putney in 1877 and educated at Bradfield College before becoming an electrical engineering student at Kensington. He later joined W. H. Allen, an engineering firm of Bedford, as a pupil. Following his military service he moved to Coventry to complete his engineering training as an 'improver' with the Daimler Company. However, his interest in electrical engineering was overtaken by his fascination for the internal combustion engine. In 1903, after leaving Daimler, he created his own firm in partnership with a Dane named Stroyer. Between them they designed and built their own car and engine, although the engines they made were much more successfully applied to other manufacturers' cars. Stroyer left after a while, leaving Lee to produce engines under the new company name — Coventry Simplex. It produced car engines for some years, with some success, including those fitted to sports cars as early as 1911. During World War 1, Coventry Simplex engines were manufactured for generating set to operate searchlights.

In 1917 Lee created Coventry Climax. The production of light car engines was running at about 400 units per week by the 1920s; indeed, car engines (and a few power units for buses) would be the mainstay for the company from 1920 until 1936, and in the 1950s and '60s the company was to gain fame as the builder of engines fitted to racing cars driven by Jack Brabham, Jim Clark, Graham Hill, Dan Gurney and Roger Penske at world-beating Grand Prix level.

From 1937 a new product was launched — a trailer-mounted fire pump. Production of these began in earnest. Coventry Climax pumps became standard equipment for the Army, Royal Air Force and ARP. Coventry Climax Type 'E' engines were also fitted to Harry Ferguson's Belfast-made tractors during wartime, and were later adapted for use in generating sets. During the war both generating sets and pumps were, of course, in great demand. Both petrol and diesel engines were pouring off the Coventry Climax assembly line. In 1945, however, the responsibility for protection against fire was transferred from national resources to local authorities, whose requirements thereafter replaced the sort of substantial orders hitherto placed by Government departments, including the military.

The founder's son, Managing Director Leonard P. Lee, arranged for eight members of staff to visit a number of leading fork truck plants in the USA. As a result, the company's first fork-lift truck — the ET 199 — was built before the end of 1946. The production model of the ET 199 was launched in 1947 after much redesigning, renamed the FTD.

The Ministry of Supply, having already placed vast orders for the company's fire pumps, was showing a keen interest. After extensive and gruelling Government-sponsored tests and trials alongside trucks from some overseas manufacturers, the MoS placed lucrative contracts for the supply of fork-lift trucks with Coventry Climax. Building fire-pumps and engines and now fork-lift trucks, Coventry Climax was as busy as it would ever be during the 1950s.

In early 1963 Coventry Climax became a part of the Jaguar Cars Ltd group of companies. By 1967 a merger with the British Motor Corporation resulted in the formation of British Motor Holdings. A year later this was followed by a further merger with the Leyland Corporation to form British Leyland Motor Corporation. The Chairman of Coventry Climax Engines was Sir William Lyons, and the Deputy Chairman and Managing Director Leonard P. Lee.

Below:
Coventry Climax was world-famous for its racing car engines, fire-pumps and fork-lift trucks. When eventually bought up by the then giant car maker British Leyland, its days were numbered. With the eventual sell-off of the various component companies belonging to the group, Coventry Climax passed to Kalmar of Sweden. Nevertheless, there are still thousands of Coventry Climax trucks operating after 20 or more years' continuous service.

A three-wheel Coventry Climax lift truck is seen at a GWR goods depot on 3 June 1947.
Ian Allan Library

Right:
Hoisting oil barrels poses no problem for this Coventry Climax fork truck, seen in 1947.
Ian Allan Library

With a choice of propulsion such as electric, battery, petrol, diesel or liquid petroleum gas (LPG), there is a truck for all situations and for economic handling of every conceivable commodity. Fruit and vegetables are moved in vast quantities in every country in the world by the mechanical muscle of fork-lift trucks on a daily basis. This Coventry Climax, fitted with a low-pressure gas cylinder, is hard at work for M. J. Morris & Sons of Hinckley, Leicestershire, in September 1978. *Ian Allan Library*

Above:
A Coventry Climax truck being employed to load box-vans with pallets of pressings on 4 August 1982. *Ian Allan Library*

Below:
Coventry Climax was chosen to supply this fork truck to move engines for the Ford Motor Co. The engines were moved by rail from engine plant to assembly plant. *Ian Allan Library*

With the break-up of British Leyland in the 1980s several divisions were sold off, including Aveling-Barford Group of Grantham, which had a long history of building road rollers, dump trucks, wheel-loaders and other construction equipment, including motor graders. The firm is currently owned by Wordsworth Holdings Ltd. Coventry Climax underwent a series of changes, eventually being bought by Swedish fork-lift truck makers Kalmar LMV, now part of the mighty Svedala organisation whose products include mining and quarrying machinery. The name 'Coventry Climax' is still visible on many fork trucks originally made by the company, some at work 20 or more years after they were built. None is currently being built under the original name.

Above:
A Coventry Climax sideloader, handling Finnish packaged softwood at Bentinck Dock in King's Lynn, Norfolk, on 14 May 1985. *Ian Allan Library*

Crompton Parkinson

Right:
Crompton Parkinson Ltd was world-famous for electric motors and battery-operated vehicles, including milk floats and electricars. An example of the latter is seen working in a warehouse on 8 October 1949.
Ian Allan Library

4 Point

Above:
One of the most unusual of all fork-lift trucks is that offered by 4 Point Lift Systems of Moline, Illinois, USA. It is itself a hybrid, but this time much more between a crane and fork-lift truck, rather than a wheel-loader and a fork-lift.

This photograph features a fork-lift-equipped machine lifting one of the company's smaller

Mobilift telescopic cranes which, in common with all other 4 Point machines, has a very low headroom, enabling it to work in factories, power stations and foundries, where very heavy loads need to be moved or placed into position by either the machine's crane or by its fork-lift attachments. *4 Point Lift Systems*

Above:
This twin-lift (having twin booms and hydraulic-lift cylinders) is able to lift objects weighing up to 50 tons. *4 Point Lift Systems*

Right:

The mighty 75-ton-capacity twin-lift from 4 Point Lift Systems is able to install very heavy presses, lathes or other very heavy machines in factories etc by using either its crane features or its fork-lift attachments, powered by a 116hp turbocharged four-cylinder diesel engine. It is mounted on four 28in-diameter, 14in-wide front wheels and two 22in-diameter, 12in-wide rear wheels which give a 90° steering angle through a planetary gear system. The machine is equipped with a telescopic counterweight section which extends to 5ft by means of two hydraulic cylinders. In standard specification with counterweight the entire machine weighs 45,000lb (20,412kg). In addition, it can use an Optional manually-positioned 10ft pinned boom extension, converting the boom length from 16ft 6in to 36ft when fully extended. (It can also be pinned at 5ft or 10ft lengths.)
4 Point Lift Systems

Gradall

A slightly less common sight in the UK is this Gradall 554 rough-terrain telescopic fork-lift truck, at work on the Channel Tunnel project at Shakespeare Cliff. Gradall is well known in the United States and Canada for its unique truck-, wheel- and crawler-mounted excavators, which use telescopic action to reach out to work, in addition to having a boom point which can rotate, aiming the bucket in any direction. *QA Photos Ltd*

Greenwood & Batley

Old-established electrical engineering firm Greenwood & Batley was one of the early manufacturers of lift trucks. The 'Green Bat' lift truck shown here was photographed on 7 December 1946.

Ian Allan Library

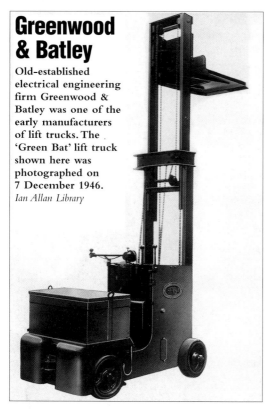

Hamech

One British company enjoying considerable success with its range of fork-lift trucks has a very interesting historical background. Hamech, as the company is known today, was founded by Mr Peter Hawkins in 1966 as an agent for Montgomerie-Reid.

Founded in 1944 by Colonel Montgomerie and Jack Reid, Montgomerie-Reid Ltd had progressed from earth-moving to the manufacture of attachments for heavy machines; upon realising the enormous potential of its home-made materials-handling machine, Montgomerie-Reid introduced the highly-versatile Wheelabout, followed by the Powerloader, Powerstack, Turnstack and Ministack. Hamech was formed from Hawkins Mechanical Handling. The company acquired the franchise to sell and service Kelvin trucks and tractors in 1971. Four years later its expansion included the UK franchise for the internationally-respected Mitsubishi range of gas and diesel trucks, and in 1980 it acquired the designs and stock-in-trade of Ransomes' fork truck division, whose origins date back to 1918 when Ransomes, Sims & Jeffries built a ride-on electric-powered truck, which astonished everyone in industry. In 1929 the Ransomes Universal electric truck was selling in vast numbers to the dockside and maritime trade, as well as to the railway goods depots. It was in 1947 that Ransomes introduced its first hydraulically-operated electric fork-lift truck. When in 1984 Hamech acquired Montgomerie-Reid, the group's experience was able to be compounded into the manufacture of over 80 models, which is what is available as we enter the new Millennium.

Below:

A Montgomerie-Reid fork-lift truck handling potatoes at a merchant's warehouse in Glasgow.
Ian Allan Library

Hunslet/West Lancs

For a unique and versatile ultra-modern four-way reach truck one need look no further than the West Lancashire Fork-lift Co, which produces the 1506, 1508, 1810 and 1612 standard models and the 3100, 3500, 4000, 4500 and 5000 with telescopic mast. Some models bear the name 'Lizard'. In warehouses where every inch of space, both on the ground and above it, has to be utilised to the full, aisles with a width range of 4ft 6in to 6ft 6in can be served adequately by the Lizard. Able to handle abnormally long loads as well as the normal pallet sizes, its unique sideloading ability is demonstrated in the accompanying photograph.

Right:
Moving heavy steel sheets is made easy with this side-loading Lizard fork-lift truck, built by Hunslet of Leeds, once famous for its range of small industrial railway engines. Although the innovative fork-lift products were transferred to West Lancs some years ago, they still bear the Hunslet name. *Author*

Hyster

The origins of the Hyster Company go back to its involvement with the logging industry of the Pacific North West in and around Oregon shortly after the turn of the century, when a blacksmith at a lumber camp built the first lumber-carrier, which consisted of four iron wheels and a load-straddling frame; with a series of cranks and levers the heavy logs were hoisted a few inches off the ground. Power was still confined to manhandling or a team of horses. In 1915, however, a similar machine was built which used electric power. By the 1920s the Willamette Iron & Steel Co was a major manufacturer of many of the machines used, though many were then powered by gasoline engines.

In February 1929 the Hyster Company was founded by Ernest G. Swigert as the Willamette-Ersted Company, manufacturing tractor-mounted winches and lumber-carriers pioneered by Willamette nine or more years earlier. The name 'Hyster' referred to the lumberjacks' cry of 'hoist here', meaning begin lifting that heavy log (once shackled to the hoisting frame).

In 1929 the Electric Steel Foundry and the Willamette Iron & Steel Co, established in Oregon, USA, formed a joint venture with a third company, Ersted Machinery, which had the factory, equipment and engineering workers to manufacture a range of handling products to serve the timber industry of the northwest USA. The Willamette-Ersted Co made logging arches and a lumber-carrier later called the Straddle Truck. It also made winches and hoisting machines for the lumber industry.

The late 1920s and early 1930s were rough times for manufacturers in the USA, but the new company survived. In 1934 it was renamed Willamette-Hyster, and around this time began experimenting in lift truck design using a tractor chassis with an aft-mounted mast and forks. A spin-off product of these early steps was the Cranemobile, later to become the Karry Krane, used around the world in the war effort. The company benefited greatly from the sales of such large numbers of these.

In 1935 the development of the BT, a fork-lift truck with a capacity of 6,000lb, was a major milestone in the company's early history. Over the first 12 months, 22 were built. Many of these were sold to lumber mills and docksides, where Willamette-Hyster had become something of a pioneer. The BT's extensive use by US and other armed forces abroad helped spread the practice to docks and warehouses around the world, and further dockside applications were rapidly discovered.

A company 'world first' soon followed with the introduction of pneumatic tyres on a large-capacity lift truck. The Jumbo, a 14,000lb-capacity truck, featured a telescopic lifting-mast, and its novel characteristics were soon incorporated into other developments such as the smaller, more manœuvrable Handy Andy.

In 1937 a Model B straddle-carrier built by Willamette Iron & Steel Co was continued by Willamette-Hyster Co, as it was renamed, actually in 1934. Various improvements were made and many new models built subsequently, including vehicles weighing 30,000lb with a capacity of 12,000lb. Fork-lift trucks such as the 3-5B/BT were built during 1935/6, utilising International Harvester petrol engines; these models were of the cushion-tyre type. These were followed in 1936-8 by the 3-6F/FT, still using International Harvester petrol engines.

By 1940 the 3-8 Handy Andy cushion-tyre lift truck had appeared, powered by a 40hp Hercules 1XB-3 four-cylinder petrol engine. What was striking about this machine was its use of a hydraulic motor, mounted so as to hoist the forks by chains through worm-type gearing, thus eliminating the need for an hydraulic cylinder which might impair the driver's vision. From 1939 to 1941 the models 3-7H/HT were built, also with International Harvester gasoline engines; in common with all the company's trucks, these were of three-wheeled type.

In 1944 the name Willamette was dropped and the company was renamed as the Hyster Co. In 1946 it built manufacturing facilities at Peoria and at Danville, both in Illinois, USA; the Danville centre was the first Hyster Co plant to be devoted exclusively to the mass production of lift trucks. In 1948 the company began to look to the international market for sales of trucks and spare parts — trucks left by the retiring military were now being pressed into civilian use. Independent dealerships were appointed by Hyster.

US sales of fork-lift trucks and ancillary services had grown from $28 million before World War 2 to $115 million by the end of the war. By 1950 the industry had a turnover of $1 billion, much of which was accounted for by Hyster's rapid expansion. In 1952, in the Dutch town of Nijmegen, the Hyster Co accomplished its biggest expansion move to date when it began to manufacture its US-designed lift trucks there.

The design of its fork-lift trucks was very important to Hyster; it was able to cater for applications in other countries outside the USA, as well as being careful to understand the requirements of both civilian and military customers. Ergonomics has long been an established manufacturing priority with many of the world's foremost producers of construction and earthmoving equipment. In this regard, the company would benefit from the services of one Henry

Below:
Hyster fork-lift trucks handling pallets of goods in and out of railway box-vans and road haulage trucks from the same loading bay. *Ian Allan Library*

Sometimes it is necessary to have specially-designed crates or pallets made to accommodate awkward or outsize loads. In this 1960s view, Ford Corsair car bodies are being loaded onto a railway flatcar at Ford's Halewood plant, using a Hyster lift truck. *Ian Allan Library*

Dreyfuss. Dreyfuss was an industrial design consultant who used the concept of human engineered products to improve operator skills and performance. He contributed greatly to the design of seat positions and the controls on Hyster fork trucks, and actually designed the Hyster 'block' logo. A great deal of attention was put to the ease of servicing the machines along with the all-important aspects of machine stability.

A particularly notable Hyster innovation was the Monomast, which made its debut in the late 1950s. This was designed so that only one mast, which contained two telescoping sections, could be used, rather than the normal two for lifting the forks. It was not universally accepted, however. Strangely, neither was the appearance of a hydraulic motor, mounted at the top of the mast, which was used on the Handy Andy of 1950, produced along with the Handy Senior from around the same time.

In 1957 the company's second European factory was opened at Hillington, near Glasgow. It was, however, transferred to Irvine shortly after. From the 1960s, manufacturing facilities were situated in Toronto, Tessenderlo (Belgium), Johannesburg, Sao Paulo and Sydney. A European marketing base was opened in London in 1962. Also in the early 1960s the company entered into an agreement with the British firm of Ransomes, Sims & Jeffries to produce Hyster-Ransomes electric fork-lift trucks in the UK. This move was superseded by the acquisition by the Hyster

Co of the Lewis Shephard Co of Massachusetts, USA, a deal which greatly enhanced Hyster's understanding of and expertise in electric motive power. William J. Frank joined the Hyster Co from the Ford Motor Co in 1966 — by which time Hyster had become a truly global company — and eventually became Company President.

In 1977 plans were unveiled for a new factory at Craigavon in Northern Ireland. This was opened in 1979 — the company's best trading year to date. However, in 1980 over 40,000 lift trucks were exported by Japan. This had a dramatic effect on sales of fork-lifts from companies such as Hyster. However, as a result of massive investment in new plant and automation at its current locations and at new facilities such as that in Berea, Kentucky, USA, over the next 10 years, with new products continually coming on stream, Hyster was able to continue as one of the world's leading manufacturers of fork-lift trucks. In 1989 Hyster was acquired by the mighty NACCO Industries Inc.

In 1991 a $31 million expansion programme was announced for the Craigavon factory, and further refurbishment of 80% of the Craigavon facility was undertaken in 1993 for the launch of a new generation of 2- to 3-tonne-capacity IC (internal-combustion-powered truck). New dealers in Germany and Italy were appointed in 1994.

Some idea of the variety of products offered by Hyster over the past 60 years is given in the following paragraphs.

From 1935 to 1942 a 12,000lb-capacity truck known as the 3-14 D/DT was produced; this heavy-duty machine was built to handle long lengths of timber and other materials, usually in the stacking yard. The 3-9J/JT Stevego built from 1939 to 1943 had a capacity of 7,500lb at 12in load centre.

Pneumatic-tyre lift trucks, including the model 4-1 QT20 fitted with a Wisconsin gasoline engine, were produced between 1943 and 1949, followed by the 4-2 QN20 from 1949 to 1956, the 4-3 QC20 from 1956 to 1961, the 4-4 H20E and H25E from 1961 to 1973, the 4-5 H30E from 1965 to the 1970s, the 4-6 YT40 from 1946 to 1957, the 4-7 (UE30,YE40, HE50) from 1956 to 1966, and 4-8 (H30F, H40F, H50F, H60F) from 1965 to 1972.

Many other models appeared during the period between 1959 and the 1970s. Interesting examples were the Jumbo (4-17) produced from 1940-2 at 15,000lb capacity, the 4-18 series (RT100, RT150) from 1941 until the early 1970s, and the 4-19 (RC150, RC160, SC180, TC200) from 1954 to 1964. The 4-20 models (H150E, H165E, H18DE, H200ES, H200E, H225E and H250E) were all in the 15,000lb to 25,000lb range, and were produced from 1965 to 1973.

The H250A and H300A in the 4-21 series from 1963 to 1973 ranged in capacity from 25,000lb to 30,000lb, and featured a choice of engines such as the Perkins 6.354 six-cylinder (rated at 130hp), Detroit Diesel 4-53N four-cylinder (140hp) or Caterpillar D320 four-cylinder (105hp) for the diesel versions, or a GMC V6-305 six-cylinder (143hp) for petrol or LPG (liquid petroleum gas).

The 4-22 range (H360A, H400A or H460A) was produced from 1959 to 1968 in capacities of between 36,000lb and 46,000lb, while the 4-23 models (H360B, H400B, H460B, H620B) produced from 1968 to the 1970s were available in capacities from 36,000lb to 62,000lb. However, from its initial manufacture in 1972 it was the 4-24 series (H700A, H800A) which stole the show for biggest and best, at capacities of 70,000lb or 80,000lb.

Hyster also produced a range of rough-terrain lift trucks which included the P40A, P50A in the 5-1 series (from 1973) and the P60A and P80A in the 5-2 range (from 1965). The 5-3 series included the RC100, produced from 1961 to 1966, and the 5-4 series of P125A, P150A, P165A, P180A, from 1964.

The straddle-truck carriers from Hyster included the 7-1 GGP (available from 1936 to 1940); 7-2 M (1940 to 1941); 7-3 M2 (1941 to 1947); 7-4 M3 (1947 to 1954); 7-5 MD (1954 to 1959); 7-6 F, FP and FNP (1936 to 1941); 7-7 MH (1941 to 1950), 7-8 MH3 (1950 to 1958); 7-9 M200E and M300E (1958 to 1960); 7-10 M200F and M300F (1960 to 1963); 7-11 M200H, M300H and M400H (1963 onwards), and 7-12 M500A (1961 to 1968) which was replaced the 7-12 M600A (currently still in production). All of these models varied from 12,000lb carrying-capacity to 60,000lb for the M600A. The 7-6 series was also able to use a fork-lift device in front of it as well as its carrying capabilities under its belly.

Hyster turret trucks were produced from 1950 to 1956, as the 8-1 2B Cargo Truck and 8-2 2C Pallet Truck. During those same years the 8-3 2D Platform

The Duncan Transrail Terminal in Salford was opened on 30 June 1983. Pride of the handling fleet is a heavy duty Hyster fork-lift truck. *Ian Allan Library*

Right:
Canterbury is the location of this haulage yard, where a Hyster fork-lift truck provides the muscle to load and unload large trailers laden with fruit or stone and aggregate. Such loads are regularly handled by A. Salvatori & Son, bound for destinations in the West Country and overseas in Belgium and the Netherlands. Seen **c1980.** *Ian Allan Library*

Left:
No matter how long, how wide or how high the load, usually there is a fork-lift to solve handling problems — all in a day's work to a Hyster truck such as this. *Ian Allan Library*

Right:
Just one of a large range of trucks from Hyster, one of the world's oldest producers of fork-lift trucks and still one of the leading manufacturers, on this occasion handling freezer cabinets for a well-known supplier of frozen foods. Hyster began making lift trucks in 1929.
Ian Allan Library

Left:
Fork-lift trucks aplenty at this London Docks facility in April 1973. All are Hyster trucks, with the exception of one Henley Husky. *Ian Allan Library*

Above:
Another view of fork trucks at work handling bulk timber from ships' cranes on board the N. R. Crump at London Docks. *Ian Allan Library*

Right:
A close-up of a heavy-duty Hyster lift truck about to move a huge pile of timber. *Ian Allan Library*

Truck was produced, as was the 8-4 2E Turretug, 8-4 2F Auto Loader and the 8-5 2A Turret Power Unit. From 1956 to 1961 the 8-6 BC Freighter and 8-7 BT Tugster units entered widespread service throughout the materials-handling industry.

The current Hyster range of trucks in all categories comprises around 100 different models, from order pickers, narrow-aisle-type machines, turret trucks, pedestrian-controlled pallet trucks, to ride-on machines like the Model E25KM (an electric-powered counter-balanced cushion-tyre lift truck), the H-series Challenger (a counter-balanced truck, sit-down operated, powered by an internal combustion engine), S-series SpaceSaver (powered sit-down lift truck), and so on.

Above:
This massive Hyster container-handling lift truck is one of a range of heavy-duty trucks, beginning with machines such as the Model H440FS, capable of lifting loads of 44,000lb (20,000kg), through a range of seven machines, to the H970-1050C with capacities of between 97,000 and 105,000lb (44,000 and 48,000kg). *Hyster Co*

Left:
A Hyster 3.20 of the XM series is one of nearly 100 models in the current Hyster range. *Hyster Europe Ltd*

Right:
These Hyster H2.00-3.20XM series lift trucks are used to handle Pepsi-Cola, and are shown here at the New Age Beverage Co warehouse in South Africa. *Hyster Europe Ltd*

Inset right:
The cab layout of the ultra-modern Hyster J1.60-2.00XMT series. Operator comfort will always result in higher productivity, whether it be in a lift truck, crane, excavator or any other item of plant. *Hyster Europe Ltd*

Right:
A Hyster 2.50 handling pallets of very heavy gunmetal ingots in a warehouse. *Hyster Europe Ltd*

Left:
This little Hyster 1.50 has all the muscle it needs to move crates of industrial liquids. *Hyster Europe Ltd*

Right:
This time it's oil barrels — again the Hyster 1.50 can cope.
Hyster Europe Ltd

Above left:
Stacking three-high is not a problem for this Hyster 3.00, even on a trailer. *Hyster Europe Ltd*

Above:
Stacking heavy-engineered products prior to loading for delivery is the task of this Hyster 1.80. *Hyster Europe Ltd*

Left:
A fork-lift truck driver will spend probably as much time driving in reverse as he will forwards. A Hyster 1.60 is seen working in a warehouse where, as in all similar situations, fork-lifts of any make have one essential ingredient: a fully-trained, highly-skilled operator, whose first thought must always be 'Safety', for himself and others around him. *Hyster Europe Ltd*

Above:

For over 15 years the family-run firm of Lowry has been redesigning and remanufacturing Hyster lift trucks, primarily for sale to customers in the USA and Canada. The Lowry range consists of the L170A (17,000lb capacity at a 24in load centre, and available with a choice of mast configurations), L190A (19,000lb at 24in, and maximum lift height of 60in), L200A (20,000lb at 24in at up to 60in) and L220A (22,000lb at 24in at up to 60in). Illustrated is a Lowry L190A built for a machinery-removal company. *Pooler-LMT Ltd*

Below:

A Lowry L220A in standard colourscheme, powered by liquid petroleum gas. *Pooler-LMT Ltd*

JCB/Chasehide

JCB has been a pioneer in many aspects of the materials-handling industry. The company is, of course, best-known for its wheel-mounted digger/loaders, now at work throughout the world, and these can operate with pallet-forks attached to the front bucket, adding to their versatility on any building site. JCB Tracked excavators, from small 1½-ton mini-diggers to large 40-ton-plus monsters, are busy on building and civil engineering sites everywhere. For several decades, the company has been involved in the manufacture of vertical-mast (and now telescopic-mast) fork-lift trucks, and also of wheel-loaders, beginning with the purchase of the Chasehide Engineering Co.

Chasehide's own machines were among the pioneers of wheel-loaders back in the days when wire ropes were used to raise and lower the buckets, before hydraulics took a stranglehold on the industry. At its works in Blackburn, the company had very successfully transformed its machines from rope-operation to hydraulics, but these, unlike Matbro's machines, were built on the rigid-frame design rather than employing articulated-frame steering, a concept which was finding more and more customers for such machines. Chasehide was acquired by JCB in 1968, whereupon production was transferred to JCB's huge Rochester complex. In a very short time Chasehide was all but forgotten, except by those employees who had brought their skills and knowledge with them from Blackburn. By the late 1970s vast numbers of JCB wheel-loaders, such as the 413 model (complete with articulated-steer) were being sold to quarries, mines, municipal organisations and to the building and civil engineering industries.

Below:
A JCB 926 RTFL (rough-terrain fork-lift truck), sporting a bucket on this occasion, turning it, albeit temporarily, into a wheel-loader. *JCB*

Above:
The JCB 930 vertical-mast rough-terrain fork-lift truck has handled thousands of tons of cement, blocks and other building materials on countless building sites since its introduction. *JCB*

Right:
The concept of a telescopic-hook fork-lift truck is one for which JCB can claim much of the credit. The introduction of its first such model — the JCB 520 — in 1977 added another dimension to rough-terrain fork-lifts which were able to work across obstacles such as trenches and manholes and still be able to reach the building under construction. This enabled work on drains and other external services to be carried out without causing disruption to building work. Only a crane would otherwise be capable of placing a pallet of bricks high on a scaffold without snuggling up tight to the foot of the scaffolding, which in many cases limited the operator's vision of what he was doing aloft. *JCB*

Above:
The JCB 520-4HL Loadall (a name previously used by JCB for its wheel-diggers/loaders of the 1950s). Such machines are now, in many cases, performing work previously carried out by farm tractors and much manual labour; the JCB is becoming as common a sight on farms as it has been on building sites. Not forgetting the range of buckets and forks available, it acts as an all-rounder down on the farm: the heavy-duty JCB Fastrac can perform all haulage work, including trailers, ploughs, harrows etc at a very fast pace. *JCB*

Right:
The JCB 525B-4HL demonstrates its reach ability in this 1985 photograph, as it places a pallet of pipes precisely where they are required on a concrete plinth. *JCB*

Right:
Where the site is well prepared for it, the vertical-mast rough-terrain fork-lift is still a very useful machine to have around to take the strain out of handling lorry-loads of building materials. This JCB 930 is a common example of such a machine. *JCB*

Left:
Is it a fork–lift truck or is it a wheel-loader? It is both. The JCB 409 has articulated steering and full four-wheel-drive — refinements customers have come to expect from JCB — but above all it is versatile. This example is equipped with pallet-forks which can rapidly be changed in favour of a bucket. *JCB*

Right:
The JCB 407B wheeled loader — a remarkable contrast from the early cumbersome machines of the 1940s and '50s. This machine has a forward-tipping excavating bucket. *JCB*

Left:
This 407B ZX wheeled loader from JCB is using a two-way tip-bucket. It can tip to the front (as is normal) or to one side — ideal for backfilling trenches or, as in this case, filling up a pavement. *JCB*

Above:
JCB certainly came up with another winner in the form of the TLT25D Teletruk, which is a fully-telescopic lift truck; although common enough among rough-terrain machines of all makes, this is a relatively new approach for smaller machines which are more suited to work in warehousing, factories, and builders' merchants' and stacking yards. This machine is fully able to reach the far side of the curtain-sided flat-bed lorry, without disturbing other pallets. *JCB*

Right:
Why move the front row of brick packs to remove or stack packs behind them? Only this type of machine could effectively achieve this, without lost time or expensive, unnecessary reshuffles of the stack. *JCB*

The Optional low cab and full free lift makes the
JCB Teletruk the ideal machine for working in
confined low-headroom applications such as this
container. There is no vertical mast to worry
about on the telescopic TLT25D. *JCB*

John Boyd

This machine, made by John Boyd in 1982, relied on winches rather than forks to perform the lifting of containers. Although much more like cranes, such machines are generally considered a hybrid of the crane and fork truck industries; while a dockside crane can load or unload containers from ships, heavy-duty fork trucks, fitted with a special container-lifting beam, can transport them around the goods terminal or container port. *Author*

Kalmar

Kalmar, a member of the Svedala group, was one of Sweden's earliest manufacturers of lift trucks. The current range includes models of up to 90 tons' capacity.

The accompanying photographs feature some of Kalmar's heavy-duty lift trucks at work.

Right:
A Kalmar container-handler on a temporary job at Arpley landfill site in Warrington. This machine, weighing 38 tons and powered by a Volvo diesel engine, could lift 30 tons on the forks; somewhat unusual was that the forks were raised entirely by hydraulic rams instead of the more common chains. *Andy Woolfenden*

Left:
This 1950s machine is seen hoisting a load of timber bound for a paper-mill. *Kalmar Industries*

Right:
Another early Kalmar fork-lift. *Kalmar Industries*

Above:
This Kalmar heavy-duty lift truck appears ready to relieve the timber-hauler of its load in one go. *Kalmar Industries*

Left:
Kalmar's LMV heavy-duty fork-lift truck. *Kalmar Industries*

Below:
One of the largest lift trucks in the Kalmar range and, possibly, the world. *Kalmar Industries*

Komatsu

Above:
Komatsu is one of the world's largest builders of heavy excavating and earthmoving equipment; indeed, it currently leads with the largest rear-end dump trucks and crawler-mounted bulldozers.

However, it also produces a comprehensive range of fork-lift trucks, of which this is one, working for a general freight haulage company in Paxton, Huntingdonshire. *Ian Allan Library*

Lansing Bagnall

The mechanical handling of goods was taken seriously by the Americans some time before companies in the UK started taking an interest. As with other things, it took only a handful of pioneers in the United Kingdom to bring the full weight of the American idea to global proportions, once the prototypes were history and large manufacturing and marketing facilities had been installed.

One of the world's foremost innovators was F. E. Bagnall, whose interests dated back to 1920 with offices at Victoria Street in London. Following visits to the Lansing Company of Lansing, Michigan, USA, Mr Bagnall, with some years of valuable experience behind him as Lansing's UK representative and distributor of

electric platform trucks, was granted the right to manufacture trucks and tractors already being produced in large numbers by the Lansing Company. He was soon able to capitalise on the large orders he was getting from the London, Midland & Scottish Railway to design and develop a small petrol-engined tractor, soon to be known as the Imp, which by 1934 had become a familiar sight at larger British railway stations. In 1937 Lansing Bagnall & Co Ltd was formed, with a nominal capital of £10,000, to acquire F. E. Bagnall's business, the first directors of the new company being Mr Bagnall and A. Forde-Nutting.

The early Lansing electric platform trucks and the Imp tractors had been built by Thomas Allen Ltd of Essex, but in 1937 the new company began to manufacture its designs. During its first year, 25 tractors and seven electric platform trucks were built. With the outbreak of war looming, this was not the most helpful fillip to the new company. In fact, by 1939 only half the number of trucks were built and the following year just 14 in total saw the light of day.

By the time Emmanuel Kaye and J. R. Sharp appeared on the scene, there were only seven employees working for the company. These two new directors were undaunted by the depressing picture, and subsequently bought the company, gaining control on 16 April 1943. Even so, the company's problems were far from over, as the Ministry of Supply announced that, following a rationalisation scheme, production would have to cease and supply of materials to the company be suspended. However, by December 1944 the embargo on the supply of materials was lifted, following an appeal by the company.

The machine tool engineering company of J. E. Shay Ltd had been employed by F. E. Bagnall (and subsequently Lansing Bagnall) from the very beginning. To assist in overcoming the current difficulties, the machine tools were transferred from its Mortlake works to the Lansing Bagnall shop at Isleworth, so that work could continue under one roof rather than two.

The Ministry of Aircraft Production had helped out with lucrative contracts for J. E. Shay Ltd for the manufacture of radio control mechanisms for bomber aircraft. This kept employees of both companies busy for a whole year. However, looking to postwar production was a vital ingredient for the future of both participants. The design and construction of prototypes for new trucks resulted in the all-new Model A towing tractor, which was to supersede the Imp. An order from the Belfast Steamship Co for 12 of the new Model H Lansing electric platform trucks (later redesignated the Model O) followed an order from the same company for six similar trucks in 1925. (Three from that contingent were still at work in Plymouth Docks in 1968.) Thereafter, innovations, new models and worldwide recognition gave rise to one of the world's largest manufacturers of fork-lift trucks. The first of these, the Model P, was the last vehicle to be designed and built at Isleworth. The first four were delivered to ICI in August 1949, followed by orders for many more, totalling 51 by the end of the year. All the production models were built at the company's new plant at Basingstoke.

Right:
An aerial view of the Lansing Bagnall works at Basingstoke. This was Number 2 Factory, seen in April 1961.
Lansing Bagnall

Left:
The first Lansing Bagnall 'Imp' tractor at London's Euston station in March 1934.
Lansing Bagnall

Left:
First staff of Lansing Bagnall to visit the Basingstoke site of the new factory. All were working at the Isleworth factory when this photograph was taken in the autumn of 1946. *Lansing Bagnall*

Above:
The LMS works at Crewe was the site of the first demonstration of the Model A tractor, seen with its operator, Mr G. A. Bull, in 1946. *Lansing Bagnall*

Left:
A demonstration of the Baker fork truck at Kodak Ltd at Wealdstone in Middlesex in 1948. *Lansing Bagnall*

Left:
The Bear Honey Company plant at Isleworth was the site of this demonstration featuring a Baker fork truck and a Model A tractor in October 1948. *Lansing Bagnall*

Right:
The Lansing Bagnall stand at the first Mechanical Handling Exhibition to be held at Olympia in 1948. *Lansing Bagnall*

Below:
The complete Lansing Bagnall range being shown off at Southampton Docks in April 1949. *Lansing Bagnall*

Below:
Lansing Bagnall's first reach truck, seen in April/May 1954, prior to its appearance at the Mechanical Handling Exhibition in June.
Lansing Bagnall

Above:
Staff in a cold store demonstration/trials of a 2-ton reach truck from Lansing Bagnall at a laboratory in Nuneaton, Warwickshire in January 1960. *Lansing Bagnall*

Below:
Lansing Bagnall TD tractor and pallet fork truck seen in the Number 1 Factory in 1955.
Lansing Bagnall

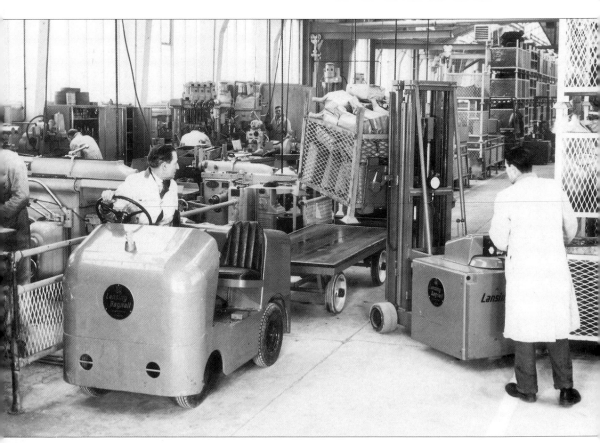

Below:
Mr F. Armitage is seen here driving a Lansing Bagnall tow tractor on a right royal occasion in October 1957, at Harlow. Passengers included HM The Queen. *Lansing Bagnall*

Above:
A low-profile Rapide fork truck supplied to the Admiralty, operating on board a ship in 1966. *Lansing Bagnall*

Below:
A TOER H tractor made by Lansing Bagnall, seen towing the Lord Mayor's Coat of Arms in Zurich, Switzerland, during the British Trade Fair in September 1963. *Lansing Bagnall*

Above:
Mr Nikita Kruschev, leader of the USSR, operating a Lansing Bagnall PP230 pallet truck at the Moscow Trade Fair in May 1961 with Mr R. S. Odd and Mr E. H. Wheeler, both of Lansing Bagnall. *Lansing Bagnall*

Trumpets blow and drums roll for the Grand Finale of the Lansing Bagnall Freight Handling Day in October 1963, but the bus taking people home is getting nowhere fast, thanks to the huge Lansing Bagnall fork-lift truck!

Right:
A pedestrian-operated pallet-truck being used at a goods depot in June 1952.
Ian Allan Library

Left:
Ride-on or pedestrian-controlled is the feature of this Lansing Bagnall pallet-truck, handling a large goods crate. The box pallet is being used by British Road Services to carry goods at depots in the Manchester area.
Ian Allan Library

Left:
A rider-controlled Lansing Bagnall truck being used at the British Road Services White City depot in Manchester on 16 April 1955. *Ian Allan Library*

Right:
Norwich Thorpe goods station on 24 May 1951, showing an example of the wide range of awkward loads being moved daily by fork-lift trucks. The Lansing Bagnall ride-on or pedestrian-operated truck would take the backache out of such moves. *Ian Allan Library*

Left:
British Railways was a major operator of pallet-trucks such as this Lansing Bagnall, seen here moving containers of ice cream on 28 February 1951. *Ian Allan Library*

Below:
Unloading a lorry full of heavy domestic appliances is made relatively easy with a mobile tail-board. Simply fit castors, a handle and a detachable tail-board and you have a trolley — the way it was done in 1953. *Ian Allan Library*

Above:
A Lansing Bagnall Rapide 3000 fork-lift truck is compact enough to drive through this box van with ease. Loading is quick and easy with a lift truck, enabling haulage contractors to achieve fast turn-arounds after every journey. *Ian Allan Library*

Left:
Mustard on the move from Colman's distribution depot, made easy with a fork-lift truck and the steel roller tracks visible on the bed of the lorry.
Ian Allan Library

You can stack high with a lift truck, using every available area of space in a warehouse.
Ian Allan Library

Lansing Bagnall's high-speed battery-electric Rapide 2000 fork trucks, operating at Express Dairies' new Nine Elms bottling centre at Battersea in London in March 1963. At the centre 480 bottles of milk were released by the conveyor plant every two minutes. Two Rapides, with their durable construction and high battery performance, were the only trucks capable of supplying the plant with empty bottles and taking filled bottles direct to waiting road vehicles (or temporary buffer storage) over a non-stop 11-hour shift, without a battery charge.
Ian Allan Library

This little pedestrian-controlled Lansing Bagnall lift truck was photographed on 24 July 1948. Many thousands of these little machines have taken the strain out of lifting and moving materials at warehouses, railway stations, factories and foundries all over the world, the design of this model having altered very little over the intervening half-century.
Ian Allan Library

Below left:
Pineapple juice from Australia. Whatever the commodity, it can always be handled a lot more easily when palletised and rolled along on a hydraulic hand-operated pallet-trolley, such as this one being used in the late 1940s at a British Railways goods depot.
Ian Allan Library

Right:
A Lansing Bagnall LPG-powered lift truck loading bottles of a popular bottled drink in China. The truck is a model FOGR 6.
Ian Allan Library

Below right:
For a few years before its acquisition by Linde, the Basingstoke-based fork truck manufacturer became known as Lansing-Henley. The Henley range of fork trucks included such models as the Husky and Hercules. This sturdy little Henley, with its twin driving wheels, has all the stability it needs to handle bundles of timber.
Ian Allan Library

Right:
Here, chocolate bars are helped along by one of two pallet-trolleys at Cadburys' Bournville factory in 1969. Note the careful strapping used to hold the cartons together on the wooden pallet.
Ian Allan Library

Right:
The Lansing-Henley Husky 10 was powered by a Ford diesel engine. It could lift 10,000lb at 24in or 7,900lb at 36in; unladen weight was 14,564lb. This particular machine is equipped with 6ft-long forks and an A-frame-type jib. *S. Harris*

Left:
The Lansing-Henley Husky was certainly a heavy-duty, sturdy lift truck. *S. Harris*

Below:
A Lansing fork truck makes very light work of handling large quantities of timber at the Port of Ipswich. *Associated British Ports*

Linde

The company was founded on 21 June 1879 and was a joint stock holding company named Gesellschaft für Linde's Eismaschinen.

In 1880 Carl von Linde began developing refrigeration machines mainly for use in breweries, slaughterhouses and ice factories. Coincidentally, other companies who were to diversify into the production of one sort of contractors' plant or another were also involved in refrigeration equipment production. In the United States, for example, the giant Manitowoc Engineering Company has become world-famous for its range of excavators and heavy-duty cranes. Winget of England produced concrete-mixers and a whole host of other plant products, including cranes. Leibherr of Germany, whose hydraulic excavators and wheel-loaders can be found on construction sites, in quarries and in open-cast mining throughout the world (along with its huge dump trucks and all types of mobile-, crawler- and tower-cranes, lifting loads of up to 1,000 tones for the top of the range), is also a major producer of refrigeration equipment.

In 1895 a patent was granted to Carl von Linde for the 'process for liquefaction of atmospheric air or other gases'. 1902 saw the development of the first air separator-employing rectification process, installed at Höllriegelskreuth (near Munich) to form the basis of Linde's own oxygen works; this process was based on a further patent owned by Carl von Linde.

In 1885 the Linde Refrigeration Company had been formed in London as the first affiliated company outside Germany. In 1907 the Linde Air Products Company was launched in Cleveland, USA, by von Linde and other partners. In 1918, the culmination of World War 1 saw the loss of numerous industrial rights, important branches and company holdings in many countries. However, despite this setback, the company forged ahead; in 1920 Linde was able to acquire the assets of Maschinenfabrik Sürth, based near Cologne, and in 1926 took over Güldner-Motoren-Gesellschaft of Aschaffenburg.

Carl von Linde died in 1934 at the age of 92. However, the sad loss of its founder had little effect on the company as a leading trader and manufacturer; indeed, four years later, in 1938, it started producing tractors at Güldner. In 1943 a manufacturing plant was set up at Schalchen, in Bavaria.

World War 2 saw the destruction of plant facilities at Höllriegelskreuth, South Mainz-Kostheim and Aschaffenburg and, once again, in 1945, the loss of industrial rights, investments and subsidiaries.

In 1955 Linde launched its first Hydrocar, a platform truck equipped with hydrostatic drive. Three years later, in 1958, celebrations were in order to mark the production of the company's 100,000th diesel engine and the launch of standard production of hydraulic units and industrial trucks (fork-lift trucks).

In 1965 the company changed its name to Linde Aktiengesellschaft, and was awarded a contract for the construction of a high-capacity petrochemical plant. Shortly afterwards, the decision was made to concentrate on growth sectors such as materials-handling equipment and hydraulics: the domestic appliances division was sold to AEG in 1967, while tractor and diesel engine production was terminated at Güldner Ashaffenburg division in 1969.

In 1973 Linde purchased the complete stock in S. E. Fahrzeugwerke GmbH, Hamburg, which has since been renamed Still GmbH, manufacturer of the Still range of fork-lift trucks and materials-handling systems. The continuance of its industrial gases section was strengthened in 1974 when new businesses were unveiled in Brazil and Australia.

Linde's lift truck business was further strengthened in 1984 through the acquisition of French fork truck manufacturer Fenwick, to be known henceforth as Fenwick-Linde S.A.R.L. In 1986 the acquisition of shares in Wagner Fordertechnik GmbH of Reuflingen took place. It was in 1989 that Lansing (Britain's own

huge fork truck manufacturer) became the subject of Linde's latest takeover, expanding even further its position in the European industrial truck market and, by coincidence, in the USA, with Baker Material Handling Corporation of South Carolina, with which Lansing had already had close ties for many years.

In 1992 a 51% interest in Fiat OM Carrelli Elevatori (fork-lift trucks to you and me) and the majority shareholding in Ciosbanc (an Italian maker of refrigeration cabinets) assured Linde's future in world markets as well as in Italy. 1993 saw a joint venture with the second-largest fork-lift truck manufacturer in China at Xiomen, while in 1994 the Juli Motorenwerk in the Czech Republic was a joint venture between Linde and Jungheinrich for the production of electric motors; Jungheinrich is another major manufacturer of fork-lift trucks.

Linde has continued its expansion in all its core businesses, with numerous acquisitions of companies involved in chemicals, liquids, refrigeration equipment, gases etc in many countries, and with many large contracts to build chemical production plants and petrochemical and fertilizer manufacturing facilities in Europe, Australia and Latin America.

Below:
A Linde H25D built in 1984. *Lisman Vorkheftrucks BV*

Matbro

One of the few large fork-truck builders in the 1950s and '60s, Matbro (Matthew Bros, pioneers of the articulated-frame wheel-loader) has continued to manufacture wheel-loaders (having acquired Bray Construction Equipment Ltd) and rough-terrain fork-lift trucks (mainly for use in the construction industry). Lancer Boss, however, took up a lot of the ideas created by Matbro and soon became number one in the heavy-fork-truck category in the UK. Boss was formed by one of Matbro's own salesmen and his brother, namely the Bowman-Shaws.

Above:
A Matbro fork-lift truck shows its rough-terrain mobility, even when carrying a huge crated generator set, in August 1950. *Ian Allan Library*

A Jones KL66 mobile crane is there to assist loading and unloading freight at this British Rail goods yard in 1980. However, the Matbro fork-lift truck is doubtless kept very busy. *Ian Allan Library*

Above:
**Matbros are again at hand to move heavy
cable-drums, in August 1950.** *Ian Allan Library*

Below:
**A 1981 Matbro Teleram, powered by an 80hp
Leyland engine driving through a Chrysler
transmission, purchased from a dealer in Cheshire
and resold to a dealer in Northern Ireland in
1996.** *Author*

Merlo

Merlo is another manufacturer which has gone from strength to strength since the launch of its first rough-terrain fork-lift in 1970 built upon the chassis of its own site-dumper. The company history began long before 1970, however. It was in 1911 that the Italian Merlo company was formed as merely a small blacksmith's shop in the town of Cúneo ('wedge') which stands between the rivers Stura and Gesso, midway between Turin and Nice.

In 1954 a new factory was opened by the company to produce machinery for foundries, paper works and the pre-fabricated concrete industry. In 1964 Mr Amilcare Merlo, his wife and sister formed a new company, 8km (5 miles) from Cúneo in the hamlet of Gervasca, to produce mobile plant. Two years later they were producing dumpers with a unique slewing skip and self-loading concrete mixers; such machines are produced by a number of Italian companies. It was at this time that their attention was drawn to a fork-lift truck (primarily for site use) built upon a dumper chassis.

In 1982 the first Merlo telescopic handlers were produced — the 3-tonne capacity SM30 series. It incorporated a unique Merlo Chassis Sideshift, which is still a feature on current models. In 1983, with the increasing popularity of Merlo, a subsidiary company was formed in France. It was 1987 that saw the introduction of the Panoramic series — the world's first telescopic handler with a side-mounted engine. In 1991 the Roto 25.11 was introduced. This also broke new ground, being the first to incorporate a slewing superstructure.

Aimed at the UK farming sector, the Turbo Farmer was made available in 1993 and, in the same year, Merlo's success in the UK was recognised with a joint-venture importer being formed (Merlo UK Ltd). In 1994 the Merlo Panoramic series won the silver and gold medals at SED (the mighty Construction Equipment Exhibition held yearly near Milton Keynes), becoming the only telescopic handler to be recognised at SED in this way. Further success was achieved two years later, the Merlo Roto and Space platform combination winning a silver medal at SED 1996.

Between 1994 and 1996 the manufacturing plant in Italy almost doubled in size from 24,000 to 40,000sq m; by the end of 1999 its size will have doubled again. Production in 1999 was in the order of 2,500 units, with an expected rise to 3,500 by the end of 2000. The Merlo product range includes three conventional handlers for site use, from 2.6-tonne to 2.8-tonne capacity (P26.6LP/SP, P28.7EVS and P28.9EVS) and handlers with sideshift, from 2.7-tonne to 6-tonne capacity (P27.9EVX to P60.10EVS — 11 models). Other products are the Roto slewing handlers: Roto 50.10EV at 5 tonnes, the 4-tonne Roto 30.13EV and a 4-tonne capacity Roto 40.21EVS (five models); and seven models of the Turbo Farmer farm handler, from the P26.6SPT/LPT at 2.6 tonnes to the 3.5-tonne P35.7EVT.

Left:
The exceptional reach of the Merlo P259EVS telescopic handler proves invaluable on building sites such as this, rendering cranes and unsafe high ladders unnecessary. *Merlo*

Right:
The Merlo range of telescopic handlers, being demonstrated in 1993. *Merlo*

Below, far right:
In conjunction with the Roto slewing telescopic handler, Merlo's Space access platform provides an ideal means of inspecting bridges, overpasses and other high structures. The combination offers the versatility of a high-reach, high-capacity, fully-revolving fork-lift truck. *Merlo*

Below right:
At one time, farmers would purchase second-hand telescopic handlers surplus to the needs of the construction industry. Nowadays manufacturers are able to offer a variety of handlers purpose-built with the farmer's needs in mind, and featuring a range of attachments such as forks, spikes, buckets, grapples and clamps. This Merlo Turbo Farmer is one of a range of seven such machines available from the Italian manufacturer. *Merlo*

Below:
Merlo was a true pioneer in developing the Roto series of fully-slewing telescopic handlers. This example is working in a stock yard for heavy construction materials, illustrating the model's exceptional reach capabilities. *Merlo*

Michigan

Left and below:
This hybrid is a Michigan 75A, built primarily as a wheel-loader. Substituting pallet-forks in place of the bucket produced a very effective and agile fork-lift of the rough-terrain type, able to operate in the most arduous ground conditions. This particular machine was a firm favourite with the Army for many years. *Author*

Moffett-Mounty

It was within the rugged terrain of the farming community of County Monaghan in the Republic of Ireland that Cecil Moffett realised his potential as a hard-working engineer who was able to make and supply the local agricultural industry with machinery found to be far too expensive from global suppliers, in an area where profits were difficult to achieve due to the poor nature of the soil. Cecil and his father soon became widely respected for their innovations, one of which included a fork-lift truck, built in 1956 by Cecil Moffett. As with all progress, things have since been transformed in this and other parts of Ireland, from the hardworking smallholdings and farms of the 1950s and '60s to the highly-mechanised and profitable farms of more recent years.

Right:
Cecil Moffett's own fork-lift truck, produced by him in 1956. *GL PR Services*

Below:
The range of Moffett-Mountys is extensive. Included are models with drive exclusive to one wheel, two or three. What is unique is its ability to handle almost every conceivable load of between 3,000lb and 6,000lb on concrete, tarmac or on rough terrain, whatever the conditions. After trucks have been loaded or unloaded, the 'Mounty' will simply attach itself to the rear of the trailer for the journey to the point of delivery. In this photograph can be seen two Moffett-Mountys in the loading area and another already attached to the rear of the huge trailer. *GL PR Services*

Right:
Demonstrating one of the range of available attachments, this Moffett-Mounty uses a bale/block clamp to handle foam used in the furniture and automotive industries. *GL PR Services*

Below:
In this instance a fire brigade has found a use for the 'Mounty'. A fork-lift can be useful in moving objects away from the seat of a fire or in providing access to a fire, perhaps in a factory. *GL PR Services*

Cecil Moffett's untimely death in October 1972 left a widow and three children — Carol (then 19), Robert (16) and Maurice (11) — and it was decided the family business should continue. Robert was the brains behind the highly successful truck-mounted fork-lift, first produced in 1986 after exhaustive tests. Only 14 were produced in the first year, but with his sister Carol chasing orders this soon grew to sales in hundreds in the ensuing years.

Other companies are now manufacturing 'Piggyback'-type fork-lift trucks, and at least one has a machine which is so compact when folded that it can fit into a box on the side of a trailer. Nevertheless, the Moffett-Mounty has become a world-beating truck which is at home handling pallets of turf, concrete blocks or bricks, timber or any one of hundreds of commodities — a fitting tribute to the late Cecil Moffett and to his children who continued his work after his death and brought the company into the 21st century.

Right:
With its ability to drive in any direction by altering the configuration of the wheels, the Moffett-Mounty can prove to be a very versatile lorry-loader, here handling long packs of profiles, which will be offloaded at the other end using a similar machine. *GL PR Services*

Above:
A closer look at the Moffett-Mounty.
GL PR Services

Above right:
Railtrack (Great Western) puts its Moffett-Mounty to good use in conjunction with Avon Fire Brigade. *GL PR Services*

Below:
The eight truck operators in this photograph have one thing in common: they all use the Moffett-Mounty to speed loading and delivery of goods.
GL PR Services

Narrow Aisle

Where narrow-aisle warehouse work is concerned, one company has for 20 years been solving complex logistics problems, enabling customers to achieve maximum storage density, handling efficiency and cost-savings. The result has been the creation of one of the world's leading manufacturers of very-narrow-aisle equipment (VNA), which is incidentally the name of the company. Its trucks, manufactured in Britain, have been sold in the Far East, North and South America, South Africa and throughout Europe.

Mr F. L. Brown did his apprenticeship with Lansing Bagnall in 1955-60. During that period he helped assemble the first reach truck Lansing Bagnall ever made. He went on in 1960 to do a postgraduate course where his thesis involved warehousing. It was then that

Mr Brown realised turret trucks would be the next step in fork truck production. He left Lansing Bagnall in 1962 to become a salesman in the Midlands for a company called Fred Myers (which among other things was involved in earthmoving equipment sales as a Caterpillar distributor for a while, and materials-handling, eventually becoming Barlow Handling). It was during the late 1960s that Mr Brown, still keen to see the emergence of turret trucks, managed to find someone to finance him, enabling him to build the first production model of this type of lift truck.

Having worked for Barlow, Brown was able to use his contacts both within and outside the company to market the concept machines throughout the world — not without the usual teething problems, not least when a Dutch dealer decided to build his own version of the turret truck.

Left:
The Narrow Aisle Combi is the ideal tool for hoisting these car bodies and placing them safely on the miles of racks. Note that the operator is at a level where he can be sure of perfect vision throughout the operation.
Narrow Aisle Ltd

Right:
Parceline has chosen Narrow Aisle's Combi for its new national distribution centre; again the operator is level with his work throughout.
Narrow Aisle Ltd

In 1977 Mr Brown formed Narrow Aisle UK Ltd. In 1982 he bought into Translift and continued making rising-cab trucks, first for Jungheinrich and then for BT Industries for 11 years, after which they decided to produce their own machines of this type.

In the early 1990s, while involved with Translift, Mr Brown developed the Bendi, with its greatly-reduced turning circle, which has proved to be very successful, gaining worldwide acceptance.

Narrow Aisle's revolutionary Rotareach was an immediate success, while its turret trucks, such as the 'man-up' Combi and Easipick order-pickers and the remarkable Flexi, all demonstrate just how the company's expertise in warehouse logistics has produced the very products for the job.

Above:
The unique configuration of Narrow Aisle's Rotareach pallet-stacker is demonstrated here at Scotia Haven Foods' warehouse. *Narrow Aisle Ltd*

Left:
A Rotareach pallet-stacker from Narrow Aisle Ltd. *Narrow Aisle Ltd*

Right:
A Narrow Aisle Combi working in racking 13m high at Brother UK in Manchester. *Narrow Aisle Ltd*

Right:
The Narrow Aisle Flexi demonstrates its articulation when manœuvring toner bins at Kodak's chemical division in Liverpool.
Narrow Aisle Ltd

Ransomes & Rapier

Ransomes & Rapier of Ipswich was a world-renowned manufacturer of steam locomotives, steam-, electric- and internal-combustion-engine-powered cranes, excavators, and (during the 1950s) some of the largest walking draglines in the world. The company was also famous for its railway turntables, water-control equipment and, as seen here, fork-lift trucks. Indeed, Ransomes was among the early entrants into the fork truck market, and may well have been the first to produce a truck that could be described as a reach truck, in the early 1900s. In the 1920s a small number of heavy-duty trucks was produced for a steel company in South Wales. All Ransomes & Rapier products were of robust and sturdy construction, and by the 1950s the company was considered by many to be the 'Rolls-Royce' of the truck industry.

Although many of Ransomes' lift trucks were plated as Ransomes & Rapier, they were built by sister company Ransomes, Sims & Jeffries, which was best-known for its vast range of agricultural implements, tractors, ride-on mowers etc. Ransomes blossomed in the 1950s and 1960s, with a large range of trucks that included its original platform trucks, tractors and pedestrian-controlled machines and counter-balance trucks from ½-ton to 3-ton capacities.

In 1961 Ransomes was bought by Hyster of the USA and its products were sold under the Hyster-Ransomes name through what is now Barlow Handling. Hyster later changed its policy and sold off Hyster-Ransomes, which was taken into the Montgomery-Reid/Lex Komatsu organisation.

Right:
This goods depot is using a variety of methods to move its cases and packages. The little 'ride-on' electric trolleys are complemented by a small Ransomes & Rapier mobile crane. *Ian Allan Library*

Left:
This large Ransomes lift truck is a Model 18/33, seen handling a British Railways box car in May 1958. *Ian Allan Library*

The Ransomes & Rapier 18/33 fork truck is seen in the late 1950s or early 1960s undergoing testing at its builder's Waterside Works in Ipswich. Visible in the background, awaiting delivery, are some of the company's other products — Rapier 410 luffing shovels.
Ransomes & Rapier Ltd

Left:
This Ransomes & Rapier 18/33 fork truck from the 1950s enables the driver to turn according to the direction of travel. This is an invaluable safety feature on such a large machine, as continually looking back over the shoulder while reversing (something fork-lifts spend much of their time doing) can cause operator fatigue and accidents.
Ransomes & Rapier Ltd

Right:
A Hyster-Ransomes reach truck is seen handling sacks of chicken feed in the Harlow Mill freight depot at British Rail's Liverpool Street station in March 1964.
Ian Allan Library

Sanderson

Roy Sanderson founded Sanderson (Fork-lifts) Ltd in 1966 to manufacture tractor three-point hitch-mounted fork-lifts for local vegetable producers; until then, the Sanderson family had been running the local blacksmith's and timber business in Croft, Lincolnshire. The first full rough-terrain fork-lift was produced in the late 1960s, based upon a Ford tractor skid unit. This was eventually developed into an 11-model product range (2,000–10,000kg lift), sold principally to farming. In early 1980, a simplified Plantman variant was developed to serve the construction and hire sectors.

In 1978 Sanderson purchased the rights to manufacture the Collins Teleporter from R. W. Collins. Ron Collins was very much a designer and continued to produce small numbers of Teleshift machines (including a novel twin-armed fork-lift) in Ledbury. The Sanderson Teleporter was first shown at the 1979 Royal Agricultural Show. After extensive re-engineering, a new, four-product range was introduced in 1980. These were 6.55m-lift-height, 2,500kg-capacity machines. The vast majority were still sold to the growing farm market. These were all rear-wheel-steer machines, which was then the norm. From 1981 to 1985 the original Teleporter product was developed into 3-tonne machines and 9m-lift, 2,500kg-capacity units.

Until 1982 export sales were almost zero, but by the end of 1984 had grown to some 35% of business. This encouraged Sanderson to purchase dumper and mixer manufacturer Winget, then in receivership. (Winget's current operations, based near Bolton, are booming, albeit under different ownership.) Sanderson followed this acquisition with that of excavator and crane maker Priestman, of Hull, from the receivers of that firm's parent company, the Acrow Group. It has often been argued that the capital and management requirements of the latter purchase were the true beginning of the decline of the Sanderson company.

In June 1985 the Teleporter 2 was introduced. This was revolutionary at the time, being the first four-wheel-steer machine designed for the farmer. Whilst the 1976 Liner Giraffe (made by one of the UK's best-remembered manufacturers of cement mixers, dumpers and other construction machinery) had four-wheel steer, it was designed for site use and could not utilise various steering modes. Sanderson had been holding a steady 19–20% share of the UK market, but this new concept increased this markedly, to over 30%. The first Teleporter 2 was a 6m-lift, 2.5-tonne-capacity machine. In 1986 it was renamed the Teleporter 622 and joined by a 7m-lift model, the 725. In late 1986, new 11m-lift models (the 1130 and 1140) were introduced with the US market in mind, and a US subsidiary was set up in late 1987.

The world economic downturn of the late 1980s led to Sanderson's being placed in receivership in November 1990. The Teleporter and Plantman parts of the business were purchased by Wordsworth Holdings soon thereafter, and production was transferred to the Aveling-Barford plant at Grantham, Lincolnshire (birthplace of Aveling-Barford steam rollers, diesel rollers, dump trucks, graders and wheel-loaders). Production of most models continued, but with little development work, and by 1996 virtually only the smaller agricultural machines were being produced.

A badging deal with German farm equipment manufacturer Claas led to that company's taking a 50% share in the Teleporter business in 1996 (the Plantman having ceased production in 1995), and gaining complete control the following year. A new factory near Bury St Edmunds is now assembling Teleporters under the Claas name.

Left:
A 1984 RWC Teleshift, an 86hp, 2.1-ton, 17ft-reach four-wheel-drive machine operating a Parmiter shear grab on a farm near Runcorn, Cheshire, in 1996. RWC, established by Ron W. Collins of Ledbury in Herefordshire, was eventually bought out by Ray Sanderson (Sanderson Fork-lifts of Lincolnshire). Mr Collins' son settled to work for Matbro in Gloucester following the demise of his father's business. *F. Fieldsend*

Right:
A Sanderson 347 TS, 3-ton, 7m, 86hp four-wheel-drive machine, built in 1983, awaiting a buyer at a plant-dealer's yard near Leeds in 1996. This one has (perhaps temporarily) swapped its forks for a bucket, showing just how versatile such machines can be. *F. Fieldsend*

Sanderson Rough-Terrain Fork-lifts from 1969:

Model	Capacity/Reach	Introduced	Replaced
SB45	2.0 tons	1969	obsolete 1978
SB55	2.6 tons	1969	1978 by SB50 and SB504
SB65	3.0 tons	1969	1978 by SB60 and SB604
SB75	3.5 tons	1969	1978 by SB70 and SB704
SB50/SB504	2.6 tons	1978	1988 by Plantman 26/264
SB60/SB604	3.0 tons	1978	1988 by Plantman 30/304
SB70/SB704	3.5 tons	1978	1991 by Plantman 40/404
SB80/SB804	4.0 tons	1978	1991 by Plantman 40/404
SB100/SB1004	5.0 tons	1978	1991 by Plantman 50/504
Plantman 26 (PM2)	2.6 tons	1988	obsolete 1994
Plantman 264 (PM2)	2.6 tons	1988	obsolete 1996
Plantman 30 (PM2)	3.0 tons	1988	obsolete 1994
Plantman 304 (PM2)	3.0 tons	1988	obsolete 1996
Plantman 40/404	4.0 tons	1991	obsolete 1994
Plantman 50/504	5.0 tons	1991	obsolete 1994
227/247	2.5 tons/7.0m	1979	1985 by 725
327/347	2.5 tons/7.0m	1979	obsolete 1985
229/249	2.5 tons/9.0 m	1979	obsolete 1985
5M26	2.6 tons/5.0 m	1986	1990/1 by 726 and 725C
622	2.25 tons/5.6m	1985	1990 by 624 and 625
623	2.3 tons/5.6m	1990	1992 by 624 and 625
725	2.5 tons/6.7m	1985	1990/1 by 726 and 725C
1130	3.0 tons/11.0m	1987	1990 by 1135
1135	3.5 tons/11.0m	1990	1993
1140 (Claas 911T)	4.0 tons/11.0m	1987 (1992)	obsolete 1994
1335	3.5 tons/13.0m	1992	obsolete Feb 1997
TX525	2.5 tons/5.3m	1992	Feb 1997 by GX525
GX525 (Claas 940GX)	2.5 tons/5.3m	1994 (1995)	Oct 1998 by 945GX
624	2.5 tons/5.6m	1992	Jan 1996 by TL6/920
625	2.5 tons/5.6m	1992	Jan 1996 by TL6/920
726 (Claas 90TT)	2.6 tons/6.7m	1990 (1993)	Jan 1996 by TL7/960/970
725c (Claas 907)	2.5 tons/6.7m	1990 (1993)	Jan 1996 by TL7/960/970
TL6 (Claas 920)	2.5 tons/5.5m	1996	Oct 1998 by 925
TL7 (Claas 976)	3.0 tons/7.0m	1996	Oct 1998 by 964
TL7/5PT 970	3.0 tons/7.0m	1996	Oct 1998 by 974
Claas 945GX	2.5 tons/5.3m	1998	(current model)
Claas 925	2.5 tons/5.5m	1998	(current model)
Claas 964	3.2 tons/6.2m	1998	(current model)
Claas 974	3.2 tons/6.2m	1998	(current model)

Above:
**A 1985 Sanderson
Teleporter 7-25 placing
pallets high on a stack.**
Claas

Above right:
**Pallet-forks and a
light-material bucket
demonstrated on
Sanderson Teleporters.**
Claas

Right:
**A Sanderson
Teleporter equipped
with a post-hole auger,
just one more job it
can perform, either on
the farm or on a
construction site.** *Claas*

Left:
**The most versatile of
all farm implements is
a telescopic handler
such as this Sanderson
Teleporter, now
produced under the
Claas name.** *Claas*

Above:
Hedge-trimmer and ditcher, both operated from a Sanderson 7.25 Teleporter from 1985. *Claas*

Left:
Using a block-clamp on a building job is a Sanderson 6.22 from 1985. *Claas*

Right:
A Sanderson 1130 Teleporter demonstrating its full 11m reach in 1987. *Claas*

Right:
A Sanderson Teleporter TL7 from 1996, when they were produced by Wordsworth Holdings of Grantham, Lincolnshire, makers of Aveling-Barford rollers, graders and dumpers. *Claas*

Left:
The 1996 Sanderson TL6 Teleporter. *Claas*

Right:
The Claas 925 Ranger with forage clamp, on agricultural duties. One of the six current models produced by Claas, it was introduced in 1998. *Claas*

Proof that a Sanderson Teleporter can dig as well as any conventional wheel-loader, at a demonstration in the late 1980s. *Dick Shelley*

With the massive expansion in aerial platforms over the past few years, it is reassuring that a Claas 975 Plus Ranger, at 3-ton capacity and 7m height, can perform the task with total confidence. *Claas*

Right:
A Sanderson 247TS Teleporter, again equipped with an excavating bucket, photographed in March 1981. *Dick Shelley*

Above:
What better way is there of stacking bales of hay or straw, than with a telescopic handler? Gone are the days when countless men handled every bale by hand or with the aid of a pitch-fork. *Dick Shelley*

Above:
A Sanderson Teleporter digging in to remove wet, sticky mud. *Dick Shelley*

Left:
The Sanderson Plantman, a vertical-mast rough-terrain fork-lift truck popular with the construction industry. *Sanderson*

Scott

Above:

A Scott three-wheel electric tractor at work at a goods depot in Birmingham, in February 1953.

Ian Allan Library

Below:

Scott Electric also made this 'rider' truck, which was being kept very busy at Bradford Forster Square goods depot in the early 1950s.

Ian Allan Library

Shelvoke & Drewry

This company, founded by Harry Shelvoke and James Drewry, has always been a specialist builder, and is perhaps best known for its refuse-collection vehicles.

However, apart from Matbro, Shelvoke & Drewry was virtually the only large fork-truck manufacturer in the 1950s and '60s, and has continued to produce 5-ton internal-combustion-engine-driven lift trucks for the heavy-industry sector.

This Shelvoke & Drewry truck is lifting a lorry which was, quite likely, also produced by S&D. The company made a name for itself with a creditable range of refuse vehicles, produced over many years. *Ian Allan Library*

Above:

This Shelvoke & Drewry Freightliner Defiant model 80/10 had a capacity of 10,000lb at 24in centre, and was equipped with a roller-mounted mast for a 17ft 6in lift. Powered by a Ford four-cylinder diesel engine (type 2701E) driving through a torque-converter transmission as standard, it had sideshifting forks and was fitted with a totally-enclosed driver's cab. The vehicle was an exhibit at the International Mechanical Handling Exhibition at Earl's Court in London from 14 to 24 May 1960. *Ian Allan Library*

Below:

A Shelvoke & Drewry Freightmaster finds lifting a large cable-drum a piece of cake, in September 1953. *Ian Allan Library*

Below:

An official British Railways inspection of a Shelvoke & Drewry Freightlifter in March 1957. *Ian Allan Library*

SkyTrak

The history of SkyTrak can be traced back to 1962 and the West Brick Buggy Company, which manufactured brick-handling equipment for the building industry. Its first products were the models 450/550 based on a Minneapolis-Moline tractor. These mimicked the function of a straight-mast fork-lift, although they lacked any form of telescopic action and featured two-wheel drive and front-wheel steer only.

Koehring PCM (Processing & Control of Materials) purchased the West Brick Buggy Company in 1966, then adding the models 700 and (in 1969) 710 to the line. The roller mast allowed for telescopic placement of loads forward of the front of the machine — creating the first fully-telescopic fork-lift truck machine. This also used that all-American favourite, the Minneapolis-Moline power-shuttle powertrain, but again employed only two-wheel drive and front-wheel steer.

In 1973 Koehring PCM introduced the Model 8034, christened the 'SkyTrak'. Being the first to adopt four-wheel drive and four-wheel steer, and also the first to use a three-section telescopic box boom, put it way ahead of its rivals. A torque-hub hydrostatic powertrain was incorporated. Other 'firsts' were its ability to tilt the frame to adjust to loading on inclines, and its use of three-mode steering.

Another development took place in 1973 when H. P. 'Hap' Mueller Jnr, then President and General Manager of Koehring PCM, left and started Badger Dynamics in Port Washington, taking several key people with him. He used the Model 8034 as the basis of the first Dynalift, but he was unsuccessful and went bankrupt a year later, in 1974. The Dynalift product line passed through several owners, ending up with Gehl in the mid-1980s. Gehl's product line included mini-excavators, skid-steer loaders and small graders.

Koehring PCM continued producing telescopic fork-lifts which included the models 4030 (from 1971), 5030 and 6034 (both from 1978). The 4030 and 5030 had a maximum lift height of 30ft 6in, while the 6034 could reach 34ft 2in. With the increased capacity these adopted the Ford power-shuttle transmission Ford diesel or petrol three- or four-cylinder engine. With two-wheel-drive with torque-hub assistance, the lockable front axle increased stability; the new range reverted to front-wheel-steer only.

In the early 1980s Koehring PCM was purchased by AMCA, which combined its new acquisition's various activities to create Koehring Construction Equipment Division. By this time, Koehring's product range included cranes and excavators (wire-rope and, later, hydraulic), compaction equipment and concrete-mixing and -placing equipment. Under AMCA ownership, Koehring CED launched further models — the 4025 in 1982 and the 5025 in 1985 — which featured an operator's cab in front of the engine, allowing the load to be placed on the driven wheels; both new models were rear-wheel-steer only, and incorporated a two-stage telescopic box boom. Other developments from Koehring CED were the 9038 (1985), 7038 (1986) and 8038 (1987). These featured higher capacities and longer reach, along with four-

Below left:
One of the first products of the West Brick Buggy Co was the Model 550 of 1962, based on a Minneapolis-Moline tractor. *SkyTrak International*

Below:
The Model 700 was introduced following Koehring's takeover of the Brick Buggy Co in 1966. *SkyTrak International*

Below:
The first model to use the SkyTrak name was the 8034, introduced by Koehring in 1973. The operator was able to enter his compartment from either side, and sat under the boom in the front of the machine, similar to that used many years ago by Taylor Jumbo on its small mobile cranes and the Coles Husky mobile crane. The rear-mounted engine acted as a counter-weight.
SkyTrak International

Below:
The Model 5522 introduced to the market by Koehring in 1985 actually was conceived by FDI in France and was imported into America to complement Koehring's own product line. It featured a fully-hydro-static drivetrain, a fully-enclosed cabin for the operator as standard equipment, and a quick-hitch attachment, enabling the operator to switch from forks to a bucket by hydraulics from the controls in the cab.
SkyTrak International

Left:
A SkyTrak Model 8042 placing a pallet of building blocks high on a building.
SkyTrak International

wheel drive and steer, and powershift mechanical-drive transmissions produced for them by Clark or Funk; power units came from Ford, John Deere, Perkins or Deutz.

A management buyout by Koehring's key managers at Port Washington created Trak International in 1987. The company was renamed SkyTrak International in 1998, and in 1999 became an autonomous division of Omniquip.

Above:
The Model 5028 was introduced by Trak International in 1991. It was originally conceived as a two-wheel-drive machine, resulting in the engine being installed at an angle so as to couple directly with the front axle. *Sky Trak International*

Above:
A SkyTrak 9038 Turbo telescopic rough-terrain fork-lift placing a pallet of blocks right where the builder needs them, at a construction site at Oak Park. *SkyTrak International*

Right:
The 6036 is one of the most popular machines in the SkyTrak range of rough-terrain fork-lift trucks. It has reach, capacity, ample drive and steering capabilities and is big enough to cope with most of the general construction work that comes its way. Introduced in 1986, the 6036 was among the last to carry the Koehring nameplate, as a management buyout by Koehring's key managers at Port Washington created Trak International in 1987. Over 8,000 Model 6036 machines have been sold to date.
SkyTrak International

Left:
Roof trusses being loaded into place by a SkyTrak 6036 using a crane jib attachment in place of forks, in another demonstration of this model's versatility.
SkyTrak International

Right:
The 42-series models (8042 and 10042) were introduced in 1992. These machines put their makers well ahead of the competition by the end of the severe recession of the early 1990s, which had brought with it many difficulties for so many companies on both sides of the Atlantic.
SkyTrak International

Left:
In 1994, a Model 10054 was unveiled. With a four-piece telescopic boom (the first of its kind) and a full 54ft boom length, it could reach a five-storey construction, displacing cranes wich had previously been the only option. *SkyTrak International*

Above:
Renamed SkyTrak International in 1998, the company celebrated with the release of the Model 3606. The first in the Millenia family of low-pivot SkyTrak machines, this machine was conceived to combine the rugged durability of a North American design with the ergonomic features of a European design. Its side-mounted engine allows for a low-profile boom to increase visibility for the operator. With even more advanced hydraulic control systems they were designed to place the load more accurately. *SkyTrak International*

Stacatruc

Stacatruc was without doubt one of the pioneers in the market and produced, in the main, counter-balance trucks in the 1- to 2-ton range. Later it was owned one third by Clark of the USA, when it was the biggest fork-truck manufacturer in the world, and a further one third was owned by BMC. This alliance was particularly good for Stacatruc during the 1960s as it thus became the sole supplier to BMC and later Leyland, and was of course able to improve its range of products from Clark. The problems with the company started when Clark bought out the other two shareholders and was itself becoming a little complacent about its competitors; the problems were compounded by the involvement of Terex, which has itself acquired crane manufacturers, hydraulic-excavator makers and dump-truck makers, and still builds its own articulated and rigid-frame dump trucks, scrapers etc. It is now one of the largest materials-handling-equipment corporations in the world.

Above:
One device designed and built for fork-lift trucks by Mr Hinder was this for picking up and emptying barrels. Besides forks, other devices which make any fork-lift truck very much more versatile are single prongs (for lifting rolls of steel, carpet or wire), scoops (for handling light materials such as chemicals) and crane beams (for lifting the likes of engines, transmissions and machine tools). Standing on his Stacatruc, Mr Hinder demonstrates his barrel-emptying device with good effect. *H. Hinder collection*

Below:
One of the early fork trucks from Clark, the company which acquired Stacatruc from the gentleman in the driving seat — Mr Harry Hinder. *H. Hinder collection*

Above:
Mr Harry Hinder was for many years one of the most knowledgeable people employed in the design and production of fork-lift trucks, having pioneered such machines as the Stacatruc, later adopted by Clark, which continued to use the Stacatruc logo on many of its products. It was Mr Hinder who introduced the Stacatruc to the Air Ministry, and he is here seen aloft with one of the Royal Air Force's senior officers beside a cargo aircraft. *H. Hinder collection*

Left:
**Stacatruc fork-lift
trucks working
alongside manually-
operated pallet trucks
and a ride-on truck.**
Ian Allan Library

Right:
**From it earliest days
Stacatruc gained a
reputation for the
ruggedness, reliability
and manœuvrability of
its products. This
machine, in use at a
British Railways goods
depot, is loading
railway vans with
foodstuffs for
distribution
throughout the
country. Stacatruc
eventually caught the
eye of giant US
fork-lift truck
manufacturer Clark,
which integrated the
British firm's products
into its own
manufacturing
programme.**
Ian Allan Library

Left and right:
**Stacatruc was
conceived by Mr Harry
Hinder, one of the
most respected names
in the development of
lift trucks in the early
years. The company
was later bought out
by Clark, which
initially retained the
Stacatruc name for use
on its own machines.**
Ian Allan Library

Still

On 1 February 1920 Hans Still established a small business repairing electric motors in Hamburg. At 22 years of age, he was full of energy and ideas, offering his customers speedy assistance with motor breakdowns, coupled with reliability and quality. A year later he embarked on the manufacture of lighting sets and portable generators; such was the success of this venture that in 1932 he set up a factory in the Berzelliusstraße in Hamburg, and by 1939 his company had more than 1,000 employees.

After the very difficult years of World War 2 Still began building industrial trucks, the first being the EK2000 2-tonne electric platform truck. In 1949 Still presented its first electronic fork-lift truck; dubbed the 'Mule', this was highly successful in its native Germany. Hans Still tragically lost his life in a car accident in 1952, but the company proved unstoppable. Today over 5,000 employees and a turnover of DM1.5 billion is testimony to its success as an innovator in the fork-lift truck market. The current company headquarters in Hamburg occupies a site covering approximately 4 square miles and produces some 12,000 electric, diesel and LPG (low-pressure gas) counterbalance trucks annually.

At its Reutlingen plant, Still manufactures warehouse trucks and handling equipment, marketed in Germany under the 'Wagner' brand-name. This range includes reach trucks, order-pickers, combination picker/stacker machines and narrow-aisle and very-narrow-aisle (VNA) equipment, plus a wide range of pedestrian-operated fork-lift trucks and pallet trucks. Still's tiller-operated equipment, stackers, order-pickers and pedestrian trucks are manufactured at Montabaire, near Paris. This was originally the Saxby manufacturing plant, and many of the machines currently produced there bear the Still-Saxby name. The Still-Saxby range includes the highly-successful, low-cost Series M-15 diesel- and LPG-powered counterbalance fork-lift trucks and the EGUX pedestrian-operated machines.

Still Materials Handling Ltd is the British subsidiary of Still GmbH. The UK operation's headquarters has been based in Bilston, West Midlands, since 1979, moving to its current premises on the site of the former Bilston Steelworks in October 1994. The UK company's turnover grew from £3 million in 1987 to £33 million in 1997.

Right:
Purchased by Linde in 1973, S. E. Fahrzeugwerke GmbH of Hamburg was later renamed Still. An R50 and an R60 are seen working for Abingdon Carpets in 1972. *Adfield-Harvey*

Below:
A whole fleet of Still lift-trucks, most of which are equipped with a special single-pronged fork for handling very heavy rolls of carpet for Abingdon Carpets, seen in 1972. *Adfield-Harvey*

Still R7 tractors at work at Ford's Dagenham plant, providing strength and stability for the movement of raw materials and machined parts in 1976. *Adfield-Harvey*

A Still R43 fork-lift truck, fitted with a special clamp enabling it to manœuvre awkward large packages/bales or goods in and out of warehouses, seen in 1979. *Adfield-Harvey*

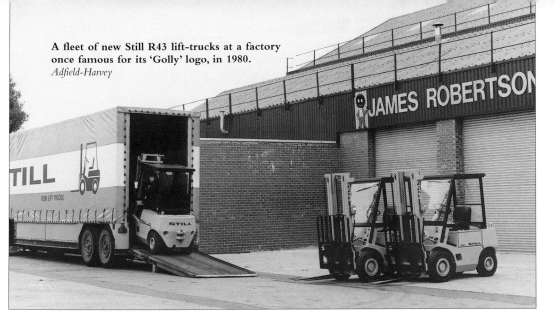

A fleet of new Still R43 lift-trucks at a factory once famous for its 'Golly' logo, in 1980.
Adfield-Harvey

Still EGU powered pallet-trucks were installed on board the latest P&O superliner, *Arcadia*, at the end of 1998. Designed for handling pallets in confined areas, the EGUs help stow the 120 tons of food and beverages required to sustain over 2,000 passengers and crew on each voyage.

Above:
The Still R60 electric fork-lift truck, as launched in 1996. It features an orthopædic seat design, an adjustable steering column and automotive-style pedal and handbrake layouts.
Adfield-Harvey

Right:
In 1998 this Still FM20 was required by Reckitt & Colman to lift heavy pallets, yet work around narrow aisles, thus optimising warehouse storage.

Right:

The Still R70, a 1.6-ton fork-lift truck, designed to run exclusively on renewable biological oils and fuel. This is the first machine of its kind, demonstrating that Still is aware of the need to conserve natural resources, and realises that oil reserves are a finite resource which will one day run out. Already a company in Lübeck has started to recycle used cooking oil from commercial catering establishments, to be used as fuel for machines and trucks, and Still is also experimenting with this process. *Adfield-Harvey*

Below.

Early in 1999, Still introduced a revolutionary new joystick control — another device which separates the machines of the distant past from the machines of the future. *Adfield-Harvey*

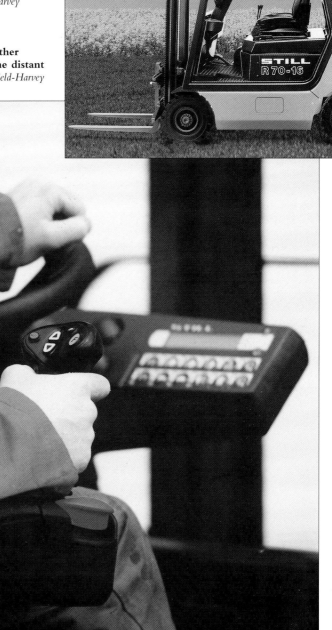

Above:

What does the future hold for fork-lift trucks? The Still RXX may give some clue, with a clear all-round-vision cab which can be raised hydraulically so that the driver always retains his visibility, even with large loads in front of him. Other features include a 'virtual' rear-view mirror which makes reversing easier and less hazardous, and an hydraulically-extending counterweight to change the centre of gravity and thus cater for differing loads. *Adfield-Harvey*

Below:

The challenge for fork-lifts of the future is to lift ever-heavier loads to ever-greater heights. Ever-more-flexible order-pickers must be the answer, such as the Still MXX, with its adjustable cab-walls that allow it to operate in particularly narrow aisles. Adjustable running-gear compensates for irregularities in the floor, and its modular construction allows for adjustable displacement of the centre of gravity. The MXX affords better all-round vision with greater convenience and comfort through separate control of the cab and load, while an angled mast gives maximum capacity to the greatest lift heights, and separate damping of the cab provides vibration-free driving. *Adfield-Harvey*

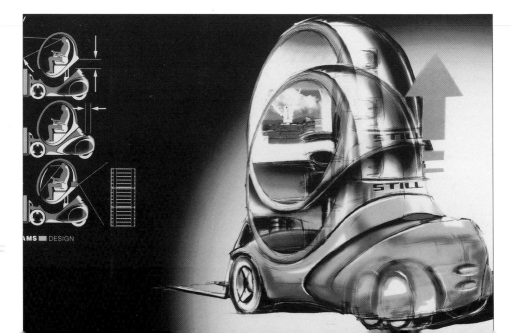

Left:
Artists' impression of the futuristic Still RXX.

Below:
Artists' impression of the futuristic Still MXX.

Right:

At present, Still is concentrating on supplying highly efficient fork-lift trucks for existing and new customers in the face of stiff competition from the many manufacturers in the field of logistics. Felix Schoeller, a producer of photographic paper, based in High Wycombe, Buckinghamshire, made a very careful evaluation of all options before choosing a fleet of 17 elect counterbalance and stacker trucks, supplied by a Still dealer on a seven-year contract. The fork-lift chosen was the Still R60-30 fitted with a Bolzomi bale clamp. Its compact dimensions and superb manœuvrability facilitate accurate positioning of paper bales and other large items. The machines in the fleet all have special rubber pads mounted on the forks to prevent damage and scarring to the floor area of the factory and storage areas, and are fitted with super-elastic non-marking tyres, so that floors need not be cleaned of tyre marks on too regular a basis. *Adfield-Harvey*

SVE

This very heavy-duty SVE truck is capable of moving hundreds of tonnes of heavy material during a normal working day. Its manœuvrability enables it to operate in confined spaces. Such machines have greatly reduced the need for cranes in recent times. *Geoff Byman*

Towmotor

It was in 1918 that Lester M. Sears conceived and began designing a straight gasoline-powered industrial towing tractor, known as the Towmotor Model A, and on 13 May 1919 Mrs Sears stepped on the starter and brought the first Towmotor tractor to life.

Designed for towing trailers loaded with motors and other engineered parts, the first model was tested at the plants of Parish & Bingham (later Midland Steel Corporation), Peerless Motor and others; initial trials proved very successful. Once production had commenced Lester M. Sears' father, a businessman, joined the Board of Directors, bringing many years of financial experience to the company's operations. He died in 1934, by which time the company was riding high despite the very real difficulties experienced in the Depression of the early 1920s.

In the early years, the Towmotor manufacturing plant was situated on Bliss Road in Euclid, Ohio. However, it was the Depression which forced them into a smaller unit (just 50ft by 100ft) on East 152nd Street in Cleveland in 1921. The Towmotor Model C first arrived on the scene on 21 October 1922. Its unique turning radius of just 5ft 3in had obvious advantages for working in warehouses, where manoeuvrability was extremely important; docks and waterfronts throughout the USA appreciated the Model C for the tight turning circle and brisk performance, backwards as well as forwards.

As the 1920s drew to a close, the Towmotor Corporation was still in business, despite the fact that its founder, Lester M. Sears, had taken a sales job with the Chilton Publishing Company at Philadelphia as its representative in Ohio. This helped to pay bills, while his evenings and weekends were spent at his drawing board, designing further Towmotor machines.

When Walter Chrysler's colossal New York skyscraper of 77 storeys was under construction, becoming the world's tallest building, and Herbert Hoover became the 31st President, money did not seem to be a problem in the United States, with an industrial worker earning an average of $25 a week and the labour unions only being able to muster 3½ million members in their early years. By 1929, however, the dark clouds were looming. On 24 October — 'Black Thursday' — over 12 million shares of stock changed hands on Wall Street. On 29 October more than 16 million shares were thrown on the market for whatever they would bring. The paper value of American common stocks had dropped $30 billion in a few weeks. It was not until the middle of 1932 that the slide in prices was arrested. The following year the banks in Cleveland, Ohio, began to open their doors again as confidence began to return.

Lester M. Sears designed and developed the first gasoline-powered fork-lift truck. The Model L was introduced in 1933 and proved to be one of the most striking advances in materials-handling since the advent

Below:
This machine is a 1936 Towmotor LT72 lift truck which came over from the USA with the armed forces during World War 2. Lovingly restored by second-year apprentice Steve Hoppe, it now carries a CAT logo. Caterpillar acquired Towmotor in 1965. *Newman Stacey*

of the crane. The very first Model L worked for its purchaser for over 20 years. It was a true pioneer in lift-truck design, incorporating under-the-load front-wheel drive and rear-wheel steering, together with hydraulic lifting and steering and high-speed forward and reverse gearing — all features which were to become the industry standard for future fork-lift trucks. In 1937 a modified version was introduced, known as the Model CL, which in 1939 evolved into the LT-46 and LT-53 (the 'T' denoting a telescopic mast).

World War 2 was the next big challenge for the Towmotor Corporation. In April 1939 President Roosevelt's attempts to obtain assurances from Hitler and Mussolini against attack on any of the 31 European and Far East nations resulted in repeated denials of aggression. However, on 1 September 1939 Germany invaded Poland, and two days later Britain and France declared war on Germany. On 7 December 1941, Japan's attack on Pearl Harbor brought the USA into what was, by now, a worldwide conflict.

Thousands of Towmotor trucks were supplied to the US armed forces, and in 1942 the Towmotor Corporation received the prestigious 'E' award and Four Service Stars in recognition of its outstanding production performance in being able to supply the Army and Navy with the equipment needed to speed the war effort. After the war, the marque's popularity continued to increase, and by 1949 the Towmotor range of lift trucks had expanded to 10 models, with capacities from 1,500 to 15,000lb, along with many attachments.

In 1951 Lester Sears handed the Presidency of the company to C. E. Smith and became Chairman of the Board. In this position he was able to guide the organisation according to the principles he had laid down as the company's founder. He was also, without doubt, one of the many leading pioneers for which the USA had long been famous. He could take his place alongside such people as J. I. (Jerome Increase) Case (founder of the mighty Case Corporation), Benjamin Holt and Daniel Best (founders of Caterpillar) and John Deere (of John Deere agricultural and construction equipment fame).

By 1952 licensing agreements were extended to manufacture Towmotor products in Australia, Denmark. The massive expansion programme had resulted in larger factories, numerous dealerships and, by 1956, acquisitions such as that of The Gerlinger Carrier Company, of Dallas, Oregon. This added heavy-capacity pneumatic fork-lift trucks and material carriers to the line. These machines had capacities of 8,000 to 40,000lb. The total Towmotor-Gerlinger product range was thus extended to 37 truck models, ranging from 1,500 to 40,000lb. This did not include heavy-duty material carriers with capacities of 12,000 to 60,000lb. Following the acquisition of Gerlinger, Towmotor became the largest single manufacturer of material-handling equipment.

Throughout the 1950s, significant strides were made in product development, with the introduction of the Streamliner series, with Models 390, 420, 460, 480P and 400P. These were trucks equipped with hydraulic brakes and larger, more powerful engines. Very high free-lift masts were incorporated into these models, if required.

It was in 1955 that engine options were offered with gasoline (petrol), diesel or LPG (liquefied petroleum gas). In 1958 Towmotor announced a major development in the transmission systems of fork trucks, with its own 'Towmostatic', based on the principle of hydrostatics. The operator was able to control the direction of the truck with a simple heel-and-toe action on a single pedal.

In 1961 the Towmotor Corporation entered into a licensing agreement with Lansing Bagnall Ltd of Basingstoke, England, for the manufacture and distribution of Towmotor products in the United Kingdom. In the same year, Towmotor acquired the Strad-o-Lift Carrier Company, which produced a tractor-hauled over-the-road trailer. Two years later, in 1963, a Cleveland, Ohio-based manufacturer of gears, sprockets, shafts and speed reduction units — The Ohio Gear Company — became an integral part of Towmotor Corporation, being henceforth known as the Ohio Gear Division. New models were introduced at an alarming rate and order books were full. Indeed, in 1964 a single order from D. H. Overmyer Warehouse Co resulted in the purchase of 500 machines, all equipped with the quad mast and capable of stacking loads to 20ft, though when collapsed the total height was just 82⅛in.

In February 1965 the Towmotor Corporation gained a listing on the New York Stock Exchange; on 9 November that same year, Towmotor Corporation became a wholly-owned subsidiary of the Caterpillar Tractor Company. In 1968 Towmotor fork-lift trucks were produced at the Caterpillar plant at Gosselies in Belgium and in Toronto in Canada (Caterpillar of Canada Ltd). From 1 October 1974 all Towmotor lift trucks displayed 'CAT' decals on the mast. By 1975 the whole product range became known as Caterpillar Lift Trucks, dispensing with the Towmotor name finally from all manufacturing facilities in July 1978.

Despite the change of identity, new models continued to stream forth, with electric versions on offer as well as those powered by internal combustion engines. Like its bulldozers, dump trucks and excavators, Caterpillar's fork-lift trucks capture a huge chunk of the market.

Right:
A Towmotor AH-52 (52,000lb-capacity) fork-lift truck built prior to the Caterpillar takeover.
Towmotor

TOWMOTOR ®

**CAPACITY
52,000 LBS.
@ 48" L.C.**
23,587 KG.
@ 1,219 mm. L.C.

Heavy duty . . . . ruggedly constructed . . . . handles 52,000 pounds (23,587 kg.)
yet operates as smoothly and easily as a compact fork lift truck.

FEATURES

Removable power package • Immediate accessibility to engine • Full 360° perimeter visibility
Envelope-type unitized frame • Interchangeability of parts • Equalized weight distribution
Fast lifting and travel speeds • Shock free, effortless hydraulic power steering
Powershift transmission with positive inching control

Left:
Looking down into a Translift Bendi, in a very narrow-aisle situation. *Translift*

Right:
Here is one example of Translift's Bendi articulated fork-lift truck, which greatly reduces the overall turning circle required when guiding its forks into the pallet. *Translift*

Valmet

This very big Valmet is the perfect tool for handling timber, using a purpose-built log-grapple. *Ian Allan Library*

Versa-Lift

Above:
The Versa-Lift is a 27-ton-capacity fork-lift truck recently introduced to the UK from the USA. It is designed primarily for the installation of heavy machinery, both outdoors and in engineering plants, chemical works etc, making use of forks and a lifting-beam, and its unique hydraulically-extendable counterweight. The 40/60 model shown is already popular amongst American operatives and is now finding favour with large British companies such as Beck & Pollitzer. *Pooler-LMT Ltd*

Below:
Another Versa-Lift 40/60, this time with extendable counterweight retracted, seen at Liverpool Docks in April 1999. This machine was supplied by Pooler-LMT to A. E. Morris Ltd of Doncaster. *Pooler-LMT Ltd*

Volvo

Right:
A Volvo articulated-steer, all-wheel-drive wheel-loader, using a specially-designed boom and grapple device for handling logs. *VME*

Left:
On this Volvo wheel-loader a more conventional fork is being utilised to move huge limestone blocks. *VME*

Yale

Right:
A Yale YCL502, one of a vast range of lift trucks from one of the world's oldest manufacturers of lift trucks and wheel-loaders. *Ian Allan Library*

Left:
This pedestrian-controlled Yale fork-lift truck has the robustness to handle cargo with all the muscle and agility of much larger trucks. Keeping goods moving is what is required in the materials-handling industry. *Ian Allan Library*

Below:
Refrigerators from Milan being moved by a Yale fork-lift truck. This particular model was the predecessor of the current line of Cushion ICE (internal combustion engine) trucks in the 3,000lb to 12,000lb-capacity range, which includes the GC-BF, GC-AF, GC-RG, GC-TG, GC-LG and GC-MG models. *Ian Allan Library*